SAFE HOME WIRING

PROJECTS

SAFE HOME WIRING
PROJECTS

Rex Cauldwell

COVER PHOTO: Susan Kahn

First printing: 1997

Printed in the United States of America

A Fine Homebuilding Book

Fine Homebuilding® is a trademark of The Taunton Press, Inc., registered in the U.S. Patent and Trademark Office.

The Taunton Press, Inc., 63 South Main Street, PO Box 5506,
Newtown, CT 06470-5506
e-mail: tp@taunton.com

Library of Congress Cataloging-in-Publication Data

Cauldwell, Rex
Safe home wiring projects / Rex Cauldwell.
p. cm.
Includes index.
ISBN 1-56158-164-X
1. Electric wiring, Interior. I. Title.
TK3285.C37 1997 97-5789
621.319'24 — dc21 CIP

To my wife, Diana, who kept the household together while I wrote. And without the proficient and dexterous help of my granddaughter, Katy, who was—and still is—demanding to draw happy faces on the computer (along with hunting Winnie the Pooh on the Internet), and without the constant help from my two cats, Little Crazy Horse and Peaches, who were always lying on the keyboard and monitor, I could have finished this book much faster.

And, finally, to my father, who taught me the hard-work ethic, and my late Uncle Bud, who taught me the trades. My Uncle Bud will always remain alive through my memories. Though the work was hard, all I remember now are the good times we had. Goodbye dear Uncle.

ACKNOWLEDGMENTS

As I mentioned in my first book, *Wiring a House,* no one person creates a book. It is the accumulation of many hands and minds. At The Taunton Press, I would like to thank Julie Trelstad, who led me through the creation process, and Tom McKenna, who led me through the editing process. I cannot imagine two better people to work with—they made the book enjoyable.

A work of this kind is very technical, and it is very easy for errors to slip by. In addition, the way houses are wired on the East Coast may not be the same on the West Coast. I would like to thank the technical readers who helped minimize these problems: Bill Goode, the toughest electrical inspector that Roanoke County, Virginia, ever had; April Elkin, of Local #637, Roanoke, Virginia; and Redwood Kardon, building inspector for the city of Oakland, California.

CONTENTS

INTRODUCTION

I am a third-generation electrician. I've been in the trades since I was a kid, helping my father and uncle, who had me doing all sorts of jobs, from fetching lunch, to carrying equipment, to snaking wires through damp, cobweb-infested crawlspaces. It was dirty, hard work, and I learned a lot from my father and uncle. Today I am a master electrician and own my own company. It wasn't what my father wanted me to do: He sent me to college so that I'd have a way out of the trades. But it didn't work. I have always loved my job, and like my uncle told me years ago, "It's not boring."

As a professional electrician, service calls are my business. You might expect most of my calls to be for emergencies or for complicated situations. Far from it. Most of my service calls have been for simple electrical installations and repairs that most anyone could handle if they only knew how to do the job properly.

But most homeowners and do-it-yourselfers don't have the knowledge to do electrical work. Well, *Safe Home Wiring Projects* gives you the practical know-how to do most any home electrical job safely and correctly.

In this book, you'll learn how to perform a visual inspection of the entire electrical system in your home. You'll get a complete list of the basic tools needed for most any electrical job. You'll see common-sense approaches to both simple installations—such as switches and receptacles—and complicated installations—such as wiring appliances. And you'll learn some of the most common mistakes that many electricians—including myself—have made so that you don't repeat them.

Safety is paramount when working with electricity. There's an old saying in the trades: "There are old electricians, and there are foolish electricians, but there are no old, foolish electricians." The majority of electrical accidents and malfunctions result from carelessness and lack of knowledge, which is why I have included numerous safety tips and warnings throughout the book.

With what you learn here, you should be able to do most of the projects in this book safely and confidently. However, some projects are more complicated (meaning dangerous) than others and so will require more experience and knowledge than you may have. One safety rule that I can't stress enough is that if you don't feel comfortable doing a job, call a professional.

1

WORKING WITH ELECTRICITY

When working with electricity, knowledge is the key to safety—ignorance literally hurts. I learned this the hard way. As a young boy, I was helping my uncle wire an old house. When I touched a wire that I should not have—a hot one—I received quite a jolt. The pain and surprise of this first electrical shock are still clearly imprinted in my memory—not a very pleasant introduction to electricity.

I have now been working as an electrician for more than 20 years and have never been seriously injured. That's not to say I haven't made painful mistakes. I have. But I haven't repeated the mistakes—I learned from them—and I've written this book to help homeowners learn from them as well.

Many homeowners are afraid to work with electricity because of the dangers involved. Their fears are justifiable, considering that electricity can maim or even kill. However, this fear prevents folks from installing receptacles, switches, and fixtures themselves, all relatively simple jobs. To accomplish these jobs, you need only a basic knowledge of electricity and a few simple tools.

ELEMENTS OF ELECTRICITY

Electricity is defined as the organized flow of electrons along a conductor. It is generated through heat, pressure, friction, light, chemical action, or magnetism. The four elements of electricity are voltage, current, power, and resistance.

The easiest way to explain these elements is to describe how electricity flows. And the simplest way to do that is to visualize a wire like a garden hose. When a hose faucet is turned on, its flowing water pushes on water already in the hose, which pushes water out the other end. Electricity works the same way. Electrons are generated and flow into a wire, which knock electrons out the other end.

The pressure that gets electricity flowing is called voltage, and it's provided by a power source, such as a battery or generator. The flow of electrons along the wire is called current, of which there are two types: direct current and alternating current (see the Glossary on p. 146). The higher the voltage, the greater the current. Both voltage and current provide power, which is the product of the voltage and current.

If you were to decrease the diameter of a garden hose, less water would be able to pass through it. As with a hose, if you were to decrease the diameter of a wire, less current could flow through it, so less electricity would reach the load. This limiting factor is called resistance—it acts like rocks in a river, trying to hold water back (the smaller the wire, the greater the resistance).

Some useful formulas

There are a few formulas that you'll find very useful when working with electricity, and they are all interrelated. Voltage, current, and resistance can be determined using Ohm's law, a principle of electricity that states that the voltage is equal to the current multiplied by the resistance ($E = IR$, where E is the voltage, I is the current, and R is the resistance).

This formula makes it easy to determine the value of one element if you know the value of the other two. For instance, if you know the resistance and the current, simply multiply the two units to find the voltage. If you know the voltage and the resistance, calculate the current using the formula $I = E/R$. If you know the current and the voltage, calculate the resistance using the formula $R = E/I$.

To calculate the power, you need to use different formulas. If you know the current and voltage, simply multiply them ($P = IE$, where P is the power). If you know only the current and resistance, use the formula $P = I^2R$. If you know the voltage and resistance, use the formula $P = E^2/R$ to calculate the power.

All these formulas are not just meant to confuse you: They have practical applications. For example, baseboard heaters are rated around 250 watts per foot. Let's say you bought a 10-ft. unit (2,500 watts). You know that the unit requires 240 volts, and you want to know how much current it will pull. Knowing the amount of current flow a specific load will pull is important because it will aid you in picking the correct size wire and breaker for the circuit.

You know the power and the voltage. To find the current in the formula $P = EI$, simply divide the power by the voltage ($I = P/E$). Plug in the numbers: 2,500 watts ÷ 240 volts = 10.42 amps. This is the current the baseboard heater will pull when 240 volts is applied to it.

TOOLS FOR WORKING WITH ELECTRICITY

To work with electricity safely, you not only need a basic knowledge of the subject, but you also need the right tools for the job. Using the right tool makes any job easier, but more important, it makes the job safer.

I admit it. I'm a tool junkie. When it comes to tools, I believe quality is everything, and I buy only the best tools available because they stand up to the rigorous workouts that I put them through. When you're shopping for tools, do some research. Buy the tool that is best suited to your needs and budget—don't just buy the cheapest one on the shelf. You'll end up paying more in the long run because it won't last as long and won't work as well as a better-quality version.

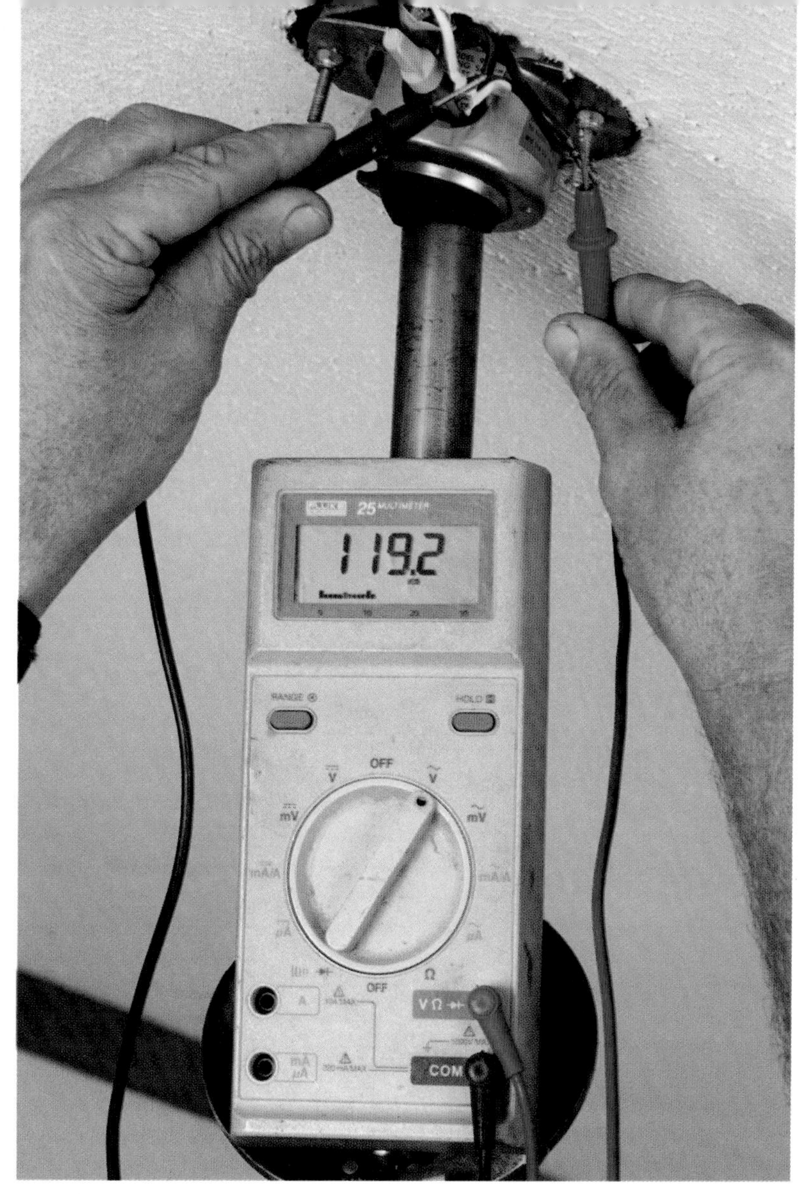

A multimeter is a must-have tool for anyone who works with electricity. It measures voltage, current, and resistance and can be used to check continuity.

Blunt-nosed side-cutting pliers are great for cutting and pulling wire.

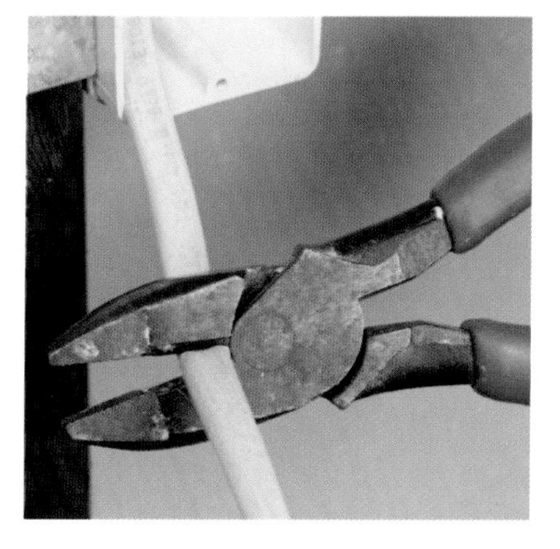

Many of the tools discussed in this section will be useful to any homeowner, and not necessarily just for electrical work. What tools you actually purchase will depend on your experience and on what kind of work you plan on doing.

Meters

A multimeter should be used as if your life depends on it—for it does. Not only can a multimeter measure voltage, current, and resistance, but it can also be used to check continuity, which verifies that two points on a circuit are electrically connected.

I have two digital multimeters (one is a backup), and I won't go on a service call without them. I use the Fluke model 25 (see the top photo at left) most often because it is autoranging (it gives me readings without my having to preset any dials). I use it for measuring voltage and resistance and for checking continuity. For checking current, though, I use my backup meter, the Fluke model 30, which is a clamp-on meter that allows me to check current without having to open the circuit. To do this, I simply clamp the two jaws around the current-carrying wire. I can check water heaters and electric baseboard heaters to verify how much current they're pulling without getting near a bare wire. The model 30 also measures voltage and resistance and can be used as a continuity checker. It is my backup meter because it is not autoranging. Both models are accurate and durable (I've dropped mine several times, and they still work).

A plug-in receptacle checker is a handy gadget for immediate analysis of the receptacle wiring. It will not only let you know if the receptacle is wired correctly, but it will also diagnose any problems with the wiring. Units for checking ground-fault circuit interrupters, called GFCIs (see the Glossary on p. 146), have

Long-nosed pliers are used primarily for bending wire into loops for insertion around screw terminals.

Diagonal-cutting pliers can cut close where blunt-nosed pliers can't fit.

a push button that simulates a ground fault on the branch circuit to see if the GFCI will trip as it should.

Hand tools

A good set of hand tools is also necessary to complete any electrical job. For safety, I recommend that you buy tools with insulated handles.

At the absolute minimum, you'll need the following: side-cutting pliers with a blunt end for cutting and pulling wire (see the bottom photo on the facing page); long-nosed pliers for bending wire ends (see the left photo above); diagonal cutters for close-in cutting (see the right photo above); wire strippers (see the middle photo at right); end-cutting pliers for cutting wire and for pulling staples (see the bottom photo at right); various short, long, fat, skinny, Phillips, and straight-bladed screwdrivers; a sharp utility knife (it might be a good idea to buy a ¼-in. and 5⁄16-in. nut driver); and some good electrical tape—I prefer 3M Super 88 (it's thick and stays put). Wood and masonry chisels, a good hammer, flat-bladed pry bars, big and small crow bars, and a hacksaw are also good to have around. To keep all these hand tools organized, buy a tool pouch or tool bucket.

Using wire strippers to remove insulation from wire is faster and safer than using a knife.

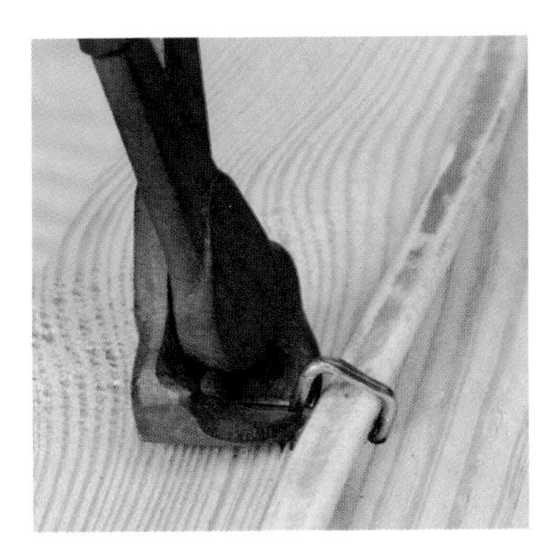

End-cutting pliers can be used for more than cutting wire. They also work well for removing staples.

An extension ladder with swivel feet provides solid support on pavement, but the feet can also be driven into the ground to anchor the ladder on a soft surface.

A reciprocating saw is one of the most versatile of the electrician's tools. It can cut through wood and nails with no problems.

A 3-in. cordless circular saw is great for cutting through wallboard. This Makita model has an adjustable blade height, and its blade makes a razor-thin, clean cut.

If you need to buy a ladder, make sure it is nonconductive (I prefer fiberglass). Look for a type 1A, which means heavy duty. When buying an extension ladder, be sure to get one that has swivel feet on the bottom. These feet provide good support on pavement, and they also cut into the ground to keep the ladder bottom from kicking out on soil (see the top photo at left).

It's also handy to have a step stool around. It is great when you need to get just a few more inches of height. My step stool is a Rubbermaid that is about 18 in. tall. It's also great for sitting on while I wire receptacles.

Power tools

Along with these assorted hand tools, a few power tools will be helpful for doing electrical work. For cutting, I recommend a reciprocating saw. A reciprocating saw will cut through anything—wood, nails, pipe, you name it (see the middle photo at left). Blades are available for cutting both metal and wood. Bimetal blades are better because they are fairly flexible—they give a bit without breaking. But be aware that bimetal blades vibrate and don't cut very straight. If you need to make straight cuts, choose extra-thick blades, which do not flex. For plugging in your power tools, buy only heavy-duty extension cords (14 gauge or heavier, with ground). Avoid the cheap, light-duty cords; they will rob your tool of valuable power.

Another handy cutting tool is a 3-in. cordless circular saw (see the bottom photo at left). I own a Makita brand, and I use it to cut through wallboard and sheathing. Its blade is extremely thin, which makes for clean cutting, and the blade height is easily adjusted. The saw is small and lightweight, so it's easy to cut straight lines, even overhead.

A cordless drill has enough power to cut most any size hole through any 2x member.

A heavy-duty right-angle drill equipped with an auger bit can cut through any wood very fast. The tool is expensive and dangerous to use because of its high torque, so the inexperienced shouldn't try one.

For drilling, buy a good-quality ⅜-in. drill. I prefer a cordless drill because it's much safer to use and is highly portable (see the left photo above). The tool is powered by a low-voltage battery, so if it malfunctions, you won't get electrocuted. It also has less torque than a corded drill, so if the bit jams in the wood, it won't break your wrist, plus you don't have to haul around an extension cord. You'll especially appreciate it when you work on a ladder or in a tight area like a crawlspace.

For heavy-duty drilling, a right-angle drill can't be beat (see the right photo above). A right-angle drill allows you to work in tight locations and gives tremendous power and torque, which make it dangerous to use. It's also very expensive, so it's more appropriate for an experienced user.

Drill bits

Drilling holes in studs, joists, and beams are common tasks for the electrician. The most efficient bit for drilling small holes is the spade bit—it can go through most anything. I've found that the bits with the two end protrusions cut faster than the standard, flat-bladed variety. Spade bits are typically available with cutting diameters of ¼ in. to 1½ in. (see the top left photo on p. 8).

Although spade bits will be sufficient for most drilling, there may be times when you need to open a large hole to pass a number of cables through. You can drill a couple of holes and cut out a circle (connecting the dots) with a reciprocating saw, but a more efficient way to open a large hole is to use a carbide-tipped hole saw. A hole saw will cut right through nails and will open a nice, clean

A spade bit with end protrusions cuts faster than the flat-bladed type.

A carbide-tipped hole saw makes quick work of opening large holes.

hole. Carbide-tipped hole saws are available in cutting diameters from ¾ in. to 6 in. (see the right photo at left).

For drilling through thick beams or logs, or for drilling through several studs at once, I prefer to use an auger bit. This bit is long and spiraled and must be used with a right-angle drill because of the torque needed to drill deep. A trick to keep the drill bit from getting stuck in deep holes is to soap the flat spiraled edges that come in contact with the

Safety Rules

When it comes to electricity, you can't be too safe. All it takes is a single, one-second mistake to kill or maim you. Here are some safety rules that will help you stay healthy.

• First and foremost, remember to turn off the power in the area in which you are working and use a multimeter to verify that it's off.

• Protect your eyes. Get safety glasses that are comfortable and scratch and fog resistant—and don't forget to wear them (see the photo below). Snipped wire ends and sparks can put out an eye or scratch it. And be sure to wear the glasses while cutting or drilling, especially overhead, where debris is falling all around your head.

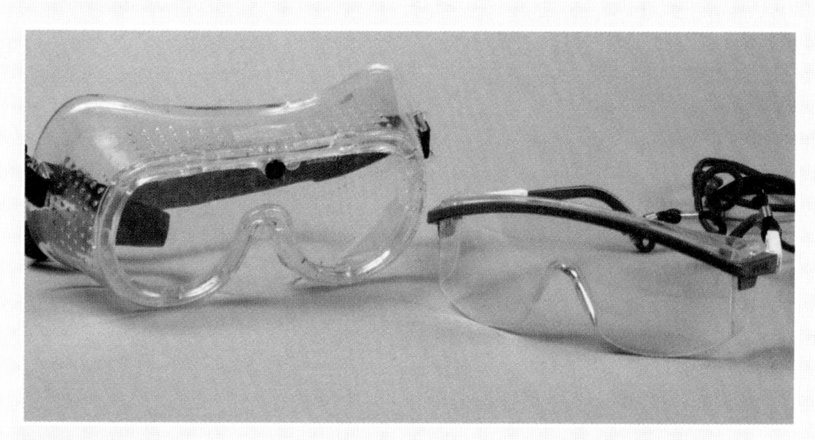

Safety glasses are a must when working around electricity and wiring. Make sure they're comfortable so that you'll wear them.

• Protect your feet, too. Drywall, main-panel covers, and other heavy objects can easily break your toes. I recommend wearing steel-toed workboots (if you don't like boots, you can buy steel-toed running shoes).

• Be sure to use the right tool for the intended job. Use screwdrivers as screwdrivers, not as pry bars or chisels. Using the right tool will make a difficult job go fast and easy; the wrong tool will make an easy job long, difficult, and dangerous.

• When working with ladders, always be aware of power lines nearby, and never use aluminum ladders, unless you have a death wish. Use only nonconductive, heavy-duty ladders.

• Electrical shock is always a possibility when working around electricity. Even though you know that you've shut off power to the room you are working in,

wood. Auger bits are available in different diameters (¼ in. to 1½ in.) and lengths (6 in., 7⅝ in., 7¾ in., and 18 in.).

For drilling holes in main panel boxes (the metal boxes that hold the circuit breakers), I use a stepped drill bit (it's really a bit shaped like a cone). One stepped drill bit can open a hole from ½ in. to over 1 in. The farther you push the bit in, the wider the hole becomes (see the photo at right).

A stepped drill bit cuts through metal easily. The farther the bit is pushed into the metal, the larger the hole it will cut.

you may have forgotten to shut off power to the room next door or upstairs. In this situation, if you cut into a wall blindly, you could easily contact a hot wire. To add an extra measure of safety, make a habit of wearing rubber gloves and shoes with rubber soles to insulate yourself from the current in case you accidentally touch a hot wire. I also recommend that you use insulated hand and power tools (or cordless tools).

• If you're working in a damp area, lay down a dry piece of plywood to stand on. The plywood will insulate you from ground and will lessen or eliminate the shock if you cut or touch a hot wire.

• Use cordless tools as much as possible when working outside, in damp areas, or on a ladder. Because cordless tools are powered by a low-voltage battery, they can't electrocute you if they malfunction. If you do use corded tools, make sure they are powered from a GFCI or plugged into a GFCI-protected extension cord.

• Cutting into walls is always dangerous—you never know what's under the surface. To be safe, always cut to the depth of the finished wallboard and no farther. For soft walls such as drywall you can use a utility knife or a drywall knife wrapped with electrical tape. I like to use my 3-in. cordless circular saw for this job because I can easily adjust it to the correct depth, and it makes a clean cut. If you use a jigsaw or reciprocating saw to cut through a wall, adjust or tilt the blade to cut only the wallboard thickness. Never cut deep and blind—you'll wind up cutting wires and plumbing.

• If you need to locate a stud to mount a switch or receptacle box, don't cut blindly into the wall. Instead, cut a small section out so you can either see or feel into the cavity. Another option is to drill a small hole, insert a bent wire into it, and swing the wire in a circle to locate the studs.

• Never drill anything while the piece is not supported securely. One time I thought I could drill through a small piece of metal while I was holding it. The bit caught the metal, and the metal turned in a circle with the bit, gouging out a very large and deep hole under my thumbnail. Please learn from my mistakes; don't repeat them.

2

INSPECTING THE ELECTRICAL SYSTEM

Like any mechanical system in your home, such as plumbing, heating, and air conditioning, the electrical system will eventually become outdated. Parts will wear down and malfunction or become damaged over the years. With the electrical system, however, malfunctioning or damaged parts can be very dangerous.

I can recall one service call I made during which the homeowner complained of receiving shocks from the gutters and downspouts. Upon close inspection, I found a worn spot on the overhead utility cable (the one that brings in power from the utility pole) that was putting voltage directly into the gutters. This was a very dangerous situation indeed. This homeowner was lucky. He called a pro, who knew what to look for and who could correct the problem.

The time to find and correct problems in the electrical system is before they become hazardous. You don't need to be an electrician to spot problem areas. Many times just a visual inspection will tell you something is wrong. The trick is knowing what to look for.

In this chapter you will learn how to perform an inspection of your home electrical system to see if there are any potential safety hazards. With what you learn here, you should be able to tell what parts need to be repaired, replaced, or upgraded. You may not be able to fix every problem yourself, but you'll know when to call a qualified electrician to help before any damage or accidents occur. (Please note that the East and West coasts have different code enforcement requirements.)

The electrical system should be inspected every couple of years. An electrical inspection should also be performed on a house you're planning to buy. The inspection begins with the service entrance, then moves on to the grounding system and the in-house wiring.

INSPECTING THE SERVICE ENTRANCE

The purpose of the service entrance is to bring power into the house. It starts at the utility transformer and terminates at the main panel.

A residence has either an aerial or buried service entrance. In an aerial service entrance, the cable installed by the utility swings from the utility transformer to the house (see the top drawing on the facing page). The cable is spliced to the service-entrance (SE) cable on the side of the house, which has been installed by an electrician (on the West Coast, the cable is run through conduit). The SE cable is attached to the meter base, which is connected to the main panel. The utility

Aerial Service Entrance

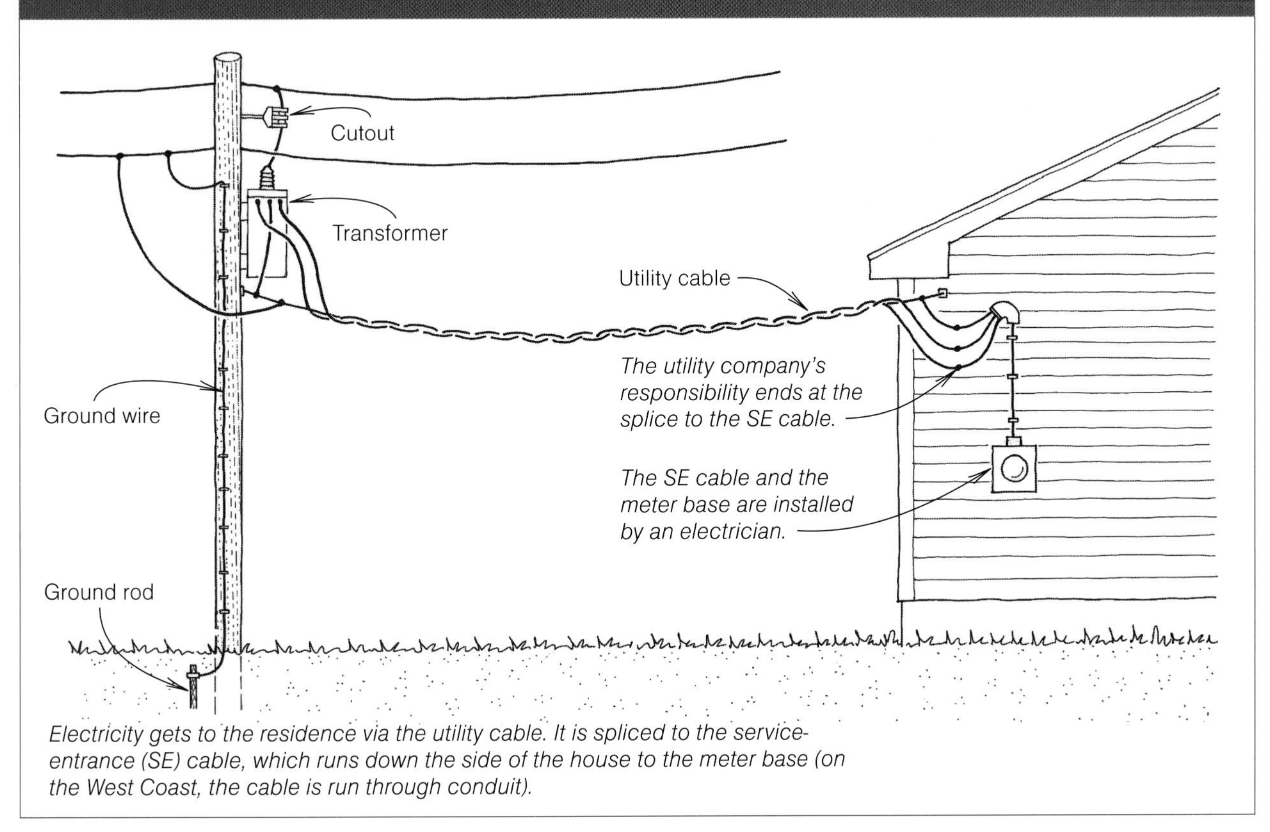

Electricity gets to the residence via the utility cable. It is spliced to the service-entrance (SE) cable, which runs down the side of the house to the meter base (on the West Coast, the cable is run through conduit).

Buried Service Entrance

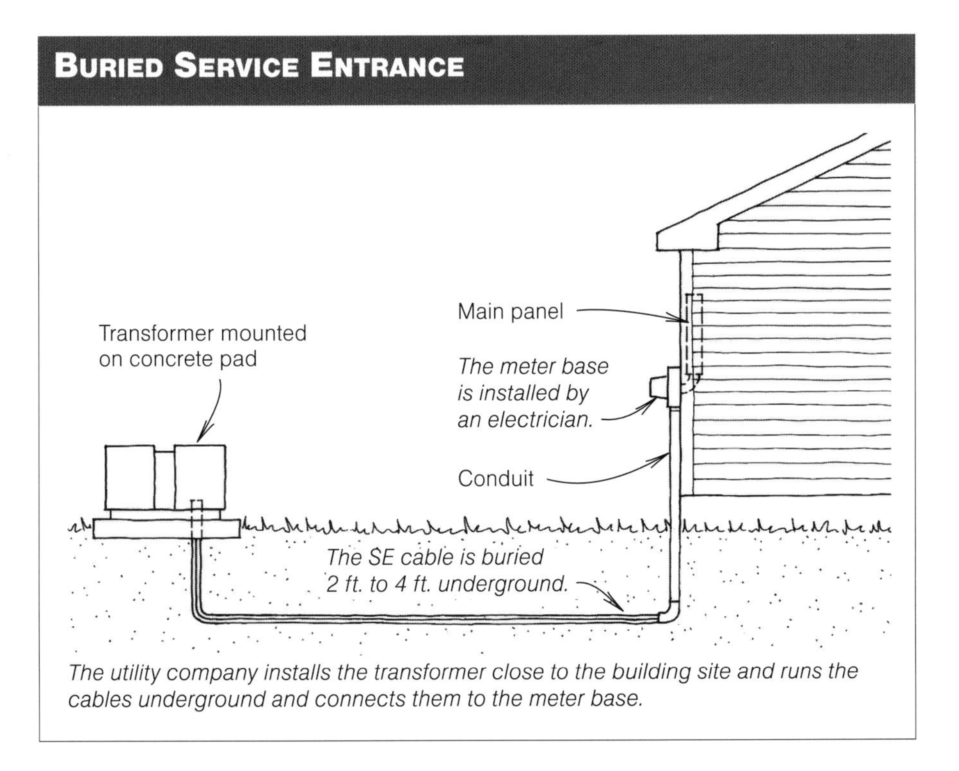

The utility company installs the transformer close to the building site and runs the cables underground and connects them to the meter base.

company's responsibility ends and the homeowner's begins at the splice to the SE cable.

In a buried service entrance, the utility company runs buried wires from a transformer to the residence and connects them directly to the meter base, which has already been installed by an electrician (see the bottom drawing on p. 11). The utility company's responsibility ends and the homeowner's begins at the meter base. (On the West Coast, the wires are run through conduit supplied by the homeowner.)

If you have an aerial service entrance, begin your inspection at the SE cable. If you have a buried service entrance, begin the inspection at the meter base.

SE cable

The SE cable sometimes will be strapped to the side of the house. Having the SE cable exposed in this manner allows the sun to shine on the cable continually. Over time, the sun's ultraviolet rays will degrade the insulation on the cable.

Visually inspect the cable from top to bottom (being careful not to hit any overhead lines with your ladder). If you see any spots where the insulation is worn or coming apart, exposing the wires inside, it needs to be replaced (see the photo below left).

Since replacement requires coordination with the utility company and the local inspection department, this job is best done by a professional electrician. I've seen SE cable so worn that its insulation was shredded or missing for its entire length.

Another common problem with SE cable strapped to the house is that the straps work loose over time. If any straps have come loose, tighten or replace them—a job any homeowner can do.

If the SE cable is run in conduit from the splice to the meter base, you won't be able to inspect the cable. In this instance, begin your inspection at the meter base.

Meter base

The inside of the meter base can only be inspected by the utility company. If an electrician thinks the meter base or the meter is bad, he should call the utility company, which will come out and open the meter base to check it.

An aerial service entrance typically creates more problems for a meter base than a buried service entrance. The enemy of any meter base is water (see the sidebar on the facing page). The SE cable enters the meter base at the top (in an aerial service) through a watertight hub. "Watertight" is actually a misnomer in this case because water almost always seeps in if the electrician who hooked up the system did not provide an additional seal with silicone rubber.

Water corrodes the aluminum terminals within the meter base, and eventually, the result will be a bad connection (it also destroys the SE cable over time). If your

The service-entrance cable should be in good condition, like the one in the top of this photo. If the cable is worn and looks like the one on the bottom, it needs to be replaced.

Meter Base with Water Damage

This meter base was destroyed by water, which wicked down the service-entrance cable and entered the base through the alleged watertight hub on top. Most experienced electricians protect against this problem by adding an additional silicon seal. This hub got its silicon a little too late.

Someone even tried to use electrical tape to stop the entrance of water (right). Eventually, the moisture destroyed the contacts inside the meter base (left). If your meter base looks like this, call a good licensed electrician immediately.

lights flicker—get brighter, then dim—the problem could be a bad connection at the meter base.

The only thing a homeowner can check on the meter base is the watertight hub on the top and the general physical condition of the meter base. If you think it looks damaged by water or anything else, call a professional electrician.

Main panel

Normally located inside the house and close to the meter base, the main panel is the heart of the electrical system. All wiring begins here and branches out to feed the circuits. It's important that everyone in the house knows where the main panel is because it is where power can be cut off to the whole house should a major electrical problem arise, such as a smoking appliance.

A main panel with circuit breakers is the modern method of providing overcurrent protection in a home.

If your home is protected by a fuse panel similar to this one, the electrical system probably needs to be upgraded.

The main panel holds either fuses or circuit breakers (officially called overcurrent protection), which protect the house wiring system from overloads, short circuits, and ground faults (see the Glossary on p. 146). If any one of these occurs, the fuse will blow, or the circuit breaker will trip, cutting off power to the circuit (see the photos above).

Checking a fuse panel If you're buying a home or own one equipped with a fuse panel, the inspection is pretty simple. The first thing to check is overfusing, which is one of the biggest problems with fuse panels. It occurs when a fuse has been installed that allows a current higher than the current rating of the wire. Fuses are made to protect to a certain amperage. For instance, a 15-amp fuse is used to protect a circuit wired with 14-gauge wire (for more on wire gauge, see the sidebar on the facing page), and a 20-amp fuse is used to protect a circuit wired with 12-gauge wire. Overfusing occurs when a fuse is replaced with one of a larger value: a 15-amp fuse protecting 14-gauge wire is replaced with a 30-amp fuse. This can allow excessive current on the circuit, which will overheat the wire, melt its insulation, and short the wires together.

For example, a 30-amp load will send a massive amount of current through both the black and white wires of a simple 120-volt circuit. The 14-gauge wires, only meant to carry a maximum of 15 amps, will begin to overheat. The insulation around the wires will degrade, usually starting at a severe bend of the wire, to a point where the wires are exposed. Once exposed, the wires could touch each other and short-circuit, which could result in a fire if the overcurrent protection is not working properly.

Overfusing was easy to do with older panels because all plug fuses had the same size base threads, which allowed a 30-amp fuse to be placed on a 15-amp circuit. But now plug fuses are designed with different types of base threads so that they cannot be interchanged (see the photo on the facing page).

To see if your panel is overfused, look at the gauge number of the cable entering the box and attached to the fuse screw. If you have a 30-amp fuse protecting 14-gauge wire, the circuit is overfused. If you can't read the writing on the cable, have a qualified electrician verify the gauge of the wire.

Wire Gauge

When referring to the gauge of a wire, the larger the number, the smaller the wire diameter (see the drawings below). For example, 12-gauge wire is larger than 18 gauge.

A large-diameter wire will have less resistance to current and will therefore be able to carry more current safely. Excessive current flowing through a small-diameter wire will overheat the wire, will damage its insulation, and could start a fire.

Small-diameter wires, such as 18 and 16 gauge, are used for low-voltage appliances. Wires of 14 and 12 gauge are used throughout the house for general-purpose circuits, such as receptacles and lighting. Ten-gauge wire is commonly used for dedicated circuits, such as 240-volt electric water heaters and dryers. Heavy-gauge, large-diameter wires are used for high-voltage applications, such as the service entrance.

Cross Sections of Copper Conductors

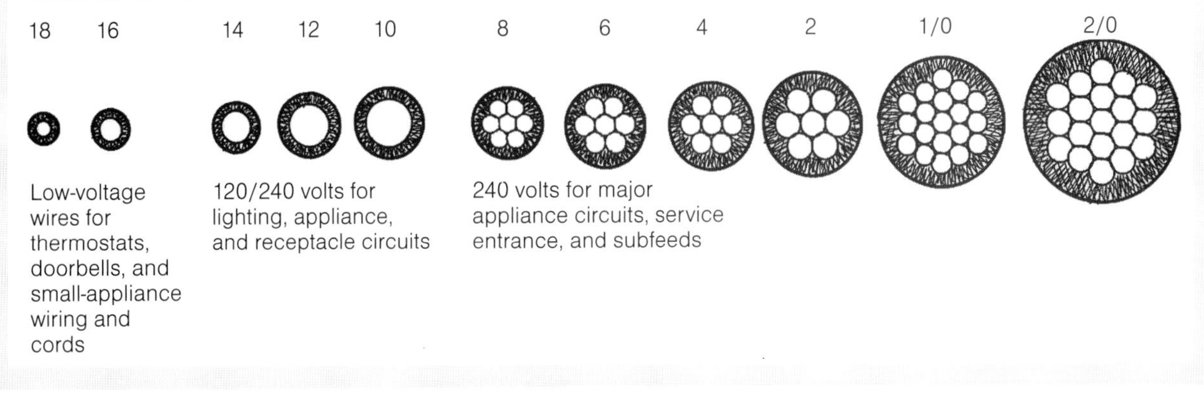

As part of the inspection, remove each plug fuse from its holder and check the center contact of the holder. If it is burned or damaged, it is no good and should not be used (see the top photo on p. 16). It's also a good indication that your system needs upgrading.

Today's plug fuses, like the two on the right, are designed with different types of base threads for different amperages, so that they cannot be interchanged as the fuse on the left.

You should also check the cartridge fuses in the panel, which provide overcurrent protection for the stove circuit and the main house (see the bottom photo on p. 16). Look for discoloration or melted spots. Cartridge fuses, when overheated, will blacken and disintegrate.

If the fuse looks okay, remove it and check its continuity to be sure that it is still good. Also check the prongs that provide the connections (once the fuse is

On the left is a burned center contact in a plug-fuse holder. A center contact should look like the one on the right. A burned contact means that something unusual happened on this circuit and that it is no longer usable.

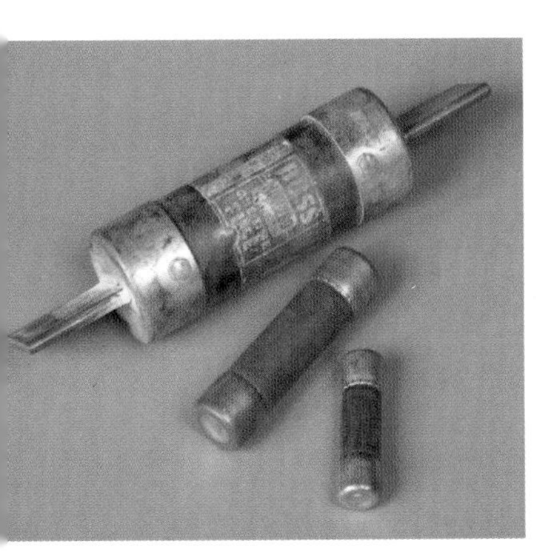

Cartridge fuses, big and small, protect circuits requiring large amounts of current.

pulled out, the prongs will no longer be hot). If they are loose, bend them back together to make the connection tighter. The fuse prongs should be a copper color with no heat discoloration (wavy, dark lines) and no burned and melted areas.

Check each cartridge fuse to be sure that no one has placed aluminum foil around the fuse to keep it from blowing due to excessive current and check behind every plug fuse to be sure no one has inserted anything, like a penny, to bypass the plug-fuse elements.

Also check the physical condition of the panel. Feel the outside of the panel box to see if it is hot. Sometimes, internal loose connections generate so much heat that they melt the box. If the box is hot, call an electrician.

It's also important to make sure you have access to the main panel. Codes dictate that nothing can be stored 3 ft. in front of the panel, from floor to ceiling. You also cannot have overhead plumbing pipes near it (these same rules apply to circuit-breaker panels, too).

In the old days, 60- or 100-amp fuse panels worked well because of the small loads placed on them, and for many houses they still work. But most of today's homes have significantly more appliances and circuits than yesterday's homes, and a 60- or 100-amp fuse panel is really not sufficient protection for the wiring. If either of these is what you have in your home, and your panel frequently blows fuses due to overloads, I'd suggest hiring an electrician and upgrading your electrical system.

Checking a circuit-breaker panel

Circuit breakers are the modern method of providing overcurrent protection in the home. They are normally very dependable, but problems arise as the panel fills to capacity and as your house electrical needs surpass the panel's size and rating. For example, you can't have a house that draws 200 amps being serviced by a panel that is only rated for 100.

The symptoms for this are pretty clear: The main breaker kicks off (disconnects the utility power from the panel) frequently under heavy loads, such as the heating and cooling system, or individual breakers frequently kick off. If you are experiencing these problems, check the panel. If it is full of breakers (no spaces left to add any), the entire panel and service entrance may need to be upgraded, which is expensive and requires a qualified electrician.

One way people get around upgrading to get more circuits on the system is to place more than one circuit on a breaker, which is a no-no. A breaker should have only one wire under each screw terminal. If you see a breaker with more than one wire under its terminal, it is probably overloaded and will kick off frequently. The solution is to add another breaker for the extra circuit. If you can't add another breaker because the panel is full, you probably need a larger panel.

One of the first things to check is that you have access to the panel (circuit-breaker panels follow the same guidelines as fuse panels). Also check to make sure that the metal electrical bus (see the Glossary on p. 146) is not exposed (see the photo above). If a tab of the bus is showing, it should be covered with a plastic snap-in insert available at electrical supply stores.

Part of the inspection involves listening to the panel. The humming sound emitted by the breakers is a good indication of their health. Low humming is okay, but if you hear very loud humming from an individual breaker, it means there's excessive current on the line (perhaps a short to ground and the breaker isn't turning off as it should). In this case throw the breaker off immediately and call a qualified, licensed electrician.

It is a safety hazard to expose the metal electrical buses of the main panel (center of photo). Cover them with plastic snap-in inserts available from electrical supply stores.

During your inspection, look for water, especially if the main panel is in the basement (this is also a problem with fuse panels). If water gets into the main panel, it will destroy everything in it, regardless of whether it is a circuit-breaker panel or a fuse panel (see the sidebar on p. 18). Water gets into a panel either from excessive dampness or from drops actually slipping down along the SE cable from the meter base.

Water Kills Circuit Breakers and Panels

Look for water, especially if the main panel is in the basement. If water gets into the main panel—either from excessive dampness or from drops slipping down along the SE cable from the meter base—it will destroy everything in the panel box.

This main breaker was destroyed by moisture, which seeped into the breaker via the stranded neutral coming from the meter base located above the main panel.

The tabs on this main-panel bus are corroded and burned from moisture. The panel was mounted directly over a drop-in range, and steam condensed on the inside, causing the aluminum to corrode and the breakers to arc.

Look for corrosion and water droplets around the breakers. Then listen for arcing (it will sound like paper tearing) and burning sounds. If you see moisture or hear such sounds, the breaker is in imminent danger of destruction and needs to be replaced immediately. If you don't replace the breaker on time, the arcing will destroy the tabs in the panel that the breakers slip into. If you find water in the panel, the first step is to stop the water from entering. Once done, take the cover off the panel and, using a large fan, blow air into the panel to dry it out. If you're lucky, you will have caught the problem before it destroyed any breakers.

Just because your panel is not in the basement, don't think that water can't get to it. I had one client whose panel was located above a drop-in range. Steam condensed on the inside of the panel and eventually corroded the aluminum and caused arcing, which destroyed the tabs on the bus.

INSPECTING THE GROUNDING SYSTEM

The grounding system connects all non-current-carrying conductors in the electrical system and then ties them to the earth via the ground rods. It protects the electrical system against ground faults, induced voltages, and voltage surges (see the Glossary on p. 146). When it comes to grounding, my rule of thumb is this: If something can become accidentally energized—through induc-

THE GROUNDING SYSTEM

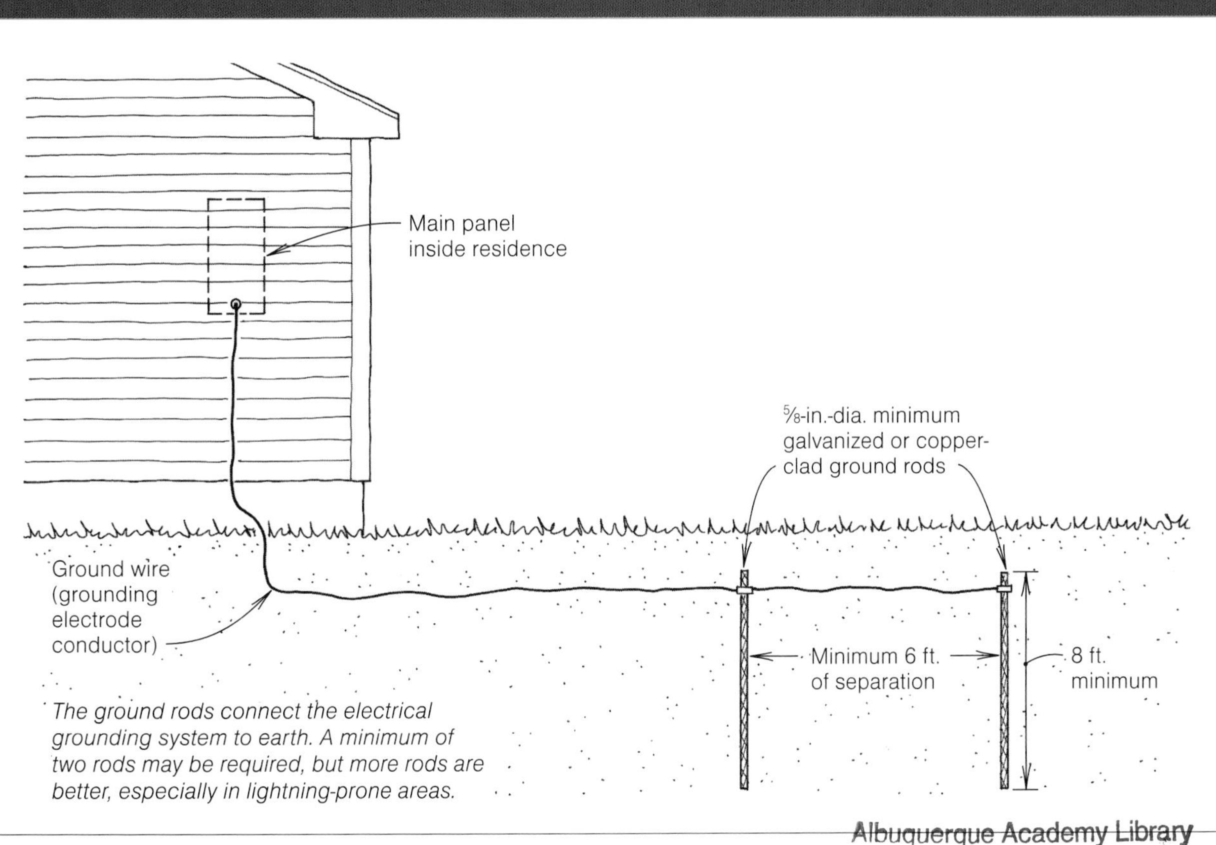

The ground rods connect the electrical grounding system to earth. A minimum of two rods may be required, but more rods are better, especially in lightning-prone areas.

tion or fault—ground it. The three things to check are the ground rods, water pipes, and receptacles.

Ground rods

The inspection for the grounding system begins with the ground rods (see the drawing on p. 19). Although they are supposed to be buried, most are not, and so they can be visually inspected. The ground rods are usually located near the main panel and sometimes near the meter base. If they are not buried, you'll be able to see them. If they are buried, try to trace the ground wire from the main panel to them.

The ground rod should be a ⅝-in.-diameter (minimum) galvanized or copper-clad rod, not some old rusty pipe. If this is what you have, replace it with the code-approved ground rod.

You should have a minimum of two ground rods (I prefer eight). However, most houses have only one because this part of the code is not strictly enforced in many parts of the country. If you have only one ground rod, and you are not experiencing problems, leave it alone. However, if you have only one, and you have problems with voltage surges or lightning strikes, you probably need to add more ground rods to the system. You may want to call a professional electrician for this job.

The wire connecting the house electrical system to the ground rods is a 4- or 6-gauge bare copper ground wire (officially called the grounding electrode conductor). Make sure it is indeed connected to the main panel. I can remember many service calls during which homeowners complained of lightning damage to appliances. When I checked the grounding system, I normally discovered a loose or cut ground wire at or near the ground rod.

Also make sure the clamp holding the wire to the ground rod is the correct type, not a hose clamp, and that it is not rusty (see the photos at left). A rusty clamp or a poor connection increases the resistance of the grounding system, making it less effective, allowing surges to jump through appliances.

If the ground wire is connected to the ground rod with a pipe clamp (right in inset), it should be replaced with the acorn clamp, which is approved for direct burial (left in inset). A pipe clamp will corrode and build up a surface resistance that can develop damaging voltage drops during lightning storms.

Grounding Metal Water Pipes

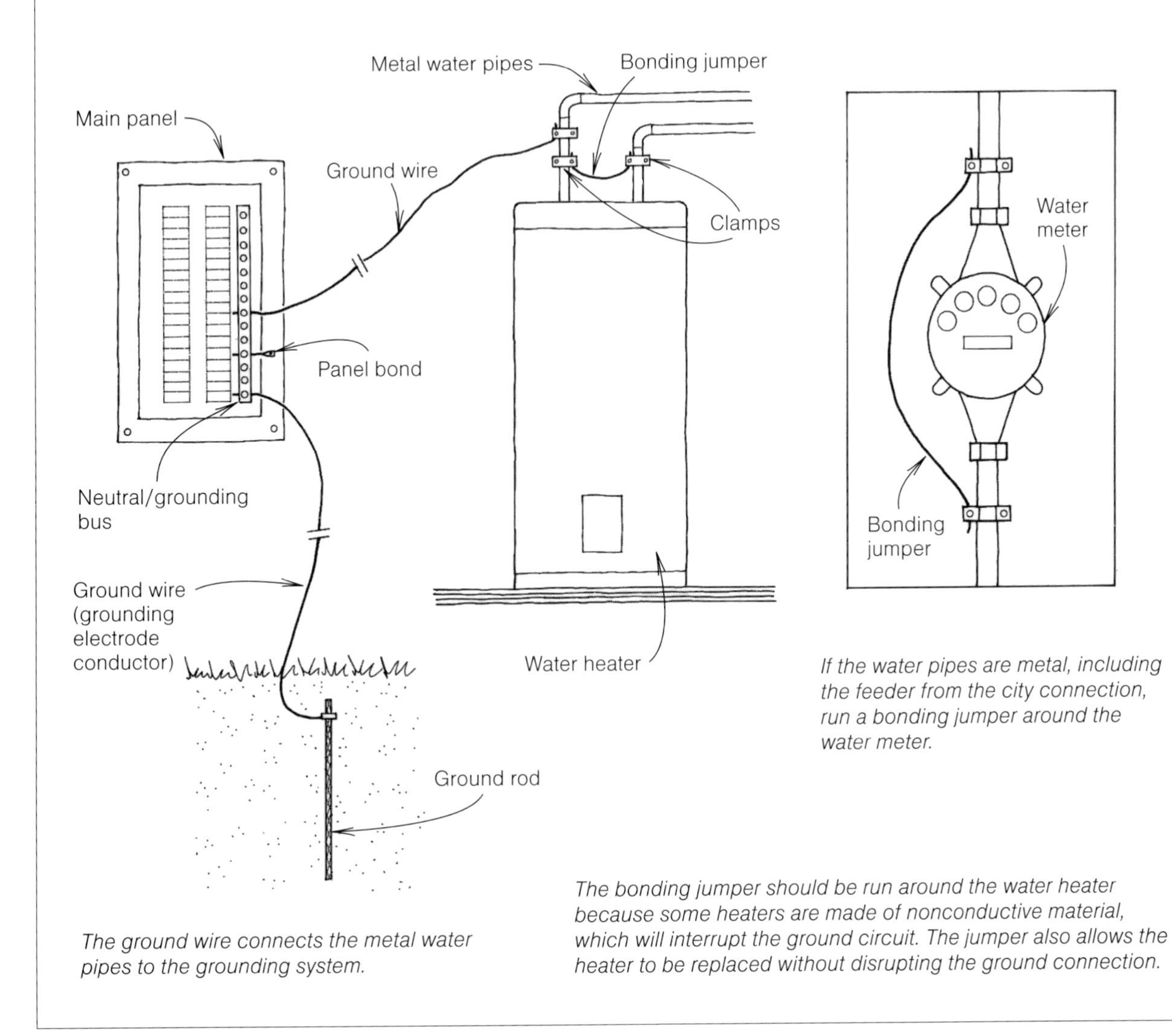

If the water pipes are metal, including the feeder from the city connection, run a bonding jumper around the water meter.

The ground wire connects the metal water pipes to the grounding system.

The bonding jumper should be run around the water heater because some heaters are made of nonconductive material, which will interrupt the ground circuit. The jumper also allows the heater to be replaced without disrupting the ground connection.

Metal water pipes

If you have metal water pipes in your house, make sure they are grounded. I made one service call where the owner was getting shocked off the copper plumbing lines, which were not grounded. I found that a hot wire had broken off a heating element and was touching the metal jacket of a water heater that was not grounded. The electricity went up the metal tank and into the metal water pipes that were screwed into the tank. The homeowner was lucky: The shock could have been fatal.

If your water pipes are metal and ungrounded, run a 4-gauge bare copper ground wire from the main panel to the pipes (see the drawings above). A bonding jumper should be placed from line to line above the water heater (many munic-

How Grounding Saves Lives

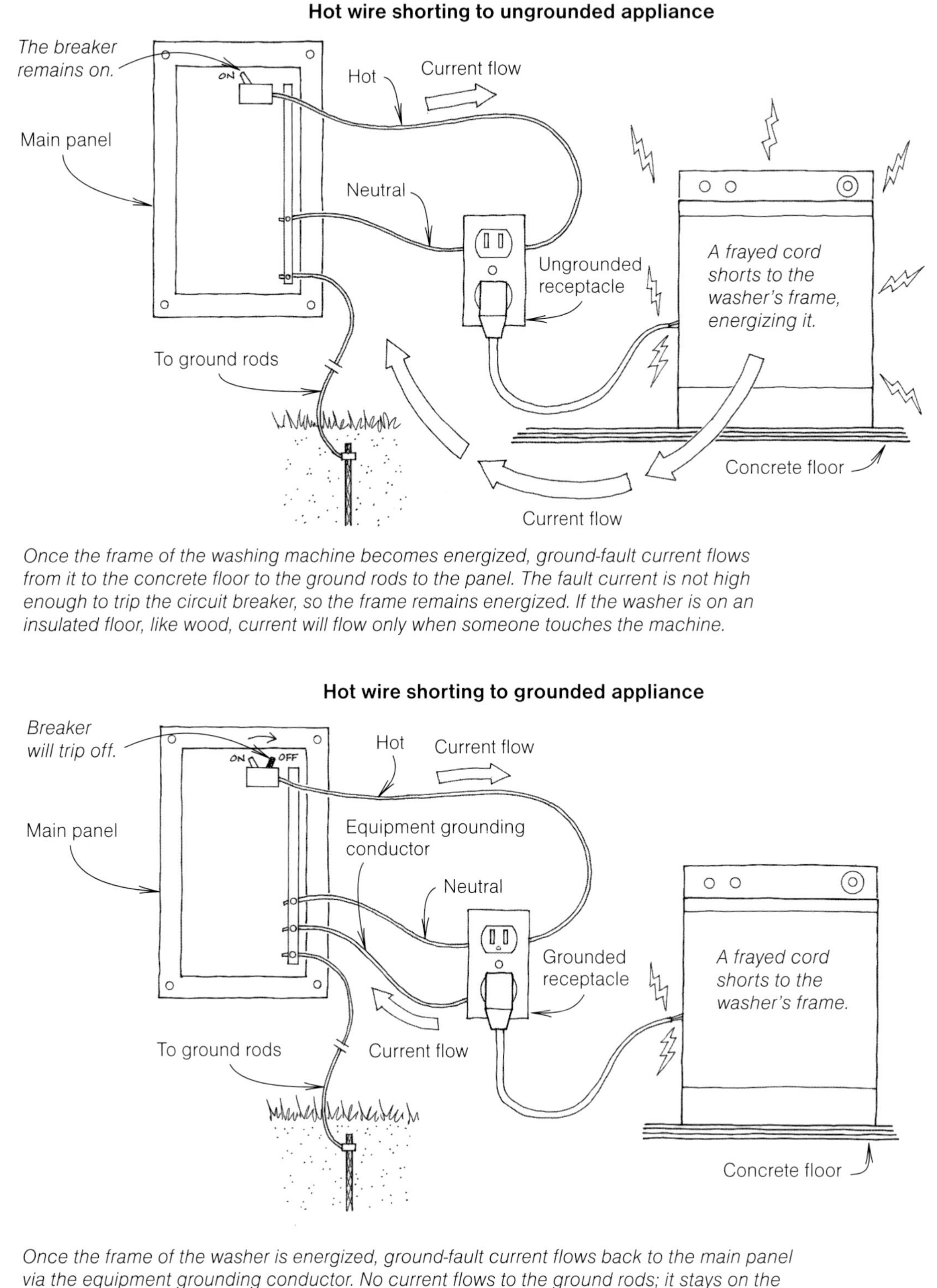

Once the frame of the washing machine becomes energized, ground-fault current flows from it to the concrete floor to the ground rods to the panel. The fault current is not high enough to trip the circuit breaker, so the frame remains energized. If the washer is on an insulated floor, like wood, current will flow only when someone touches the machine.

Once the frame of the washer is energized, ground-fault current flows back to the main panel via the equipment grounding conductor. No current flows to the ground rods; it stays on the wiring and trips the circuit breaker.

ipalities require the gas line to a gas heater to be bonded as well). The bonding jumper continues the ground connection of the pipes even if the water heater is nonmetallic or if it is disconnected or replaced. If you have city water and all metal pipes out to the utility, a bonding jumper should be placed around the water meter for the same reasons.

(Other items that should be grounded are garage-door rails, metal ductwork, and metal beams. Simply clamp a ground wire to these items and connect it to the main panel.)

Receptacles

During your inspection, make sure all receptacles are grounded. In older houses, you can see instantly whether the receptacles are ungrounded: The receptacles have no grounding slots (the grounding slot is the one above or below the two narrower slots). Ungrounded receptacles are unacceptable by today's standards.

Most appliances need to be grounded as required by their manufacturer. And if you think an adapter that allows a three-pronged appliance to be plugged into a two-pronged receptacle will serve as a ground, you're wrong. These adapters, what I call 2-3 cheater plugs, do not ground appliances unless they are connected to a metal conduit system. They only allow you to plug the appliance in. (A grounded appliance has its metal frame connected to the main panel's grounding system via the grounding slot on the receptacle.) Not grounding an appliance that requires it is a code violation and a safety hazard. If the hot wire touches the appliance's metal frame, and the frame is not grounded, the frame becomes hot and can electrocute anyone who touches it (see the drawings on the facing page).

To rectify this problem, there are three options: running a separate, bare ground wire (the same gauge as the hot wire) from the main panel's grounding bus (see the Glossary on p. 146) to any appliance that needs to be grounded; installing a new, grounded circuit where needed; or rewiring the house completely with modern cable, which has an equipment grounding conductor—a bare copper wire commonly known as the ground wire. Never connect a ground wire from the receptacle ground screw to metal pipes in an attempt to obtain a ground through the plumbing. There is no assurance that the plumbing is connected to earth and no guarantee that it will remain so if it is. If you have any doubt about what needs to be done or how to do it, call a professional electrician.

One other note: Just because you see three-slotted receptacles in the house, don't assume that they're grounded or grounded correctly. There are two ways to be sure a receptacle is grounded: Test it with a plug-in tester or remove the receptacle and check to see if the ground wire does indeed exist.

INSPECTING THE IN-HOUSE WIRING

In this part of the inspection, you should be looking for faulty wiring of receptacles, damaged receptacles and switches, undersized wiring, and damaged wiring. Because receptacles and switches are often wired on the same circuit, any problems along one connection could spell trouble for everything else on the circuit (for more on wiring receptacles and switches, see Chapter 3).

Faulty wiring

Faulty wiring is both an electrical problem and a fire hazard. I've seen circuits arc and throw sparks on the carpet, I've seen receptacles and switches burn, and I've also seen various appliances go up in smoke. Sometimes it takes electrical testing to determine faulty wiring; other times it's just simple observation.

The first thing to check is the number of receptacles in the house. A typical mistake is wiring to minimum code, which is an easy error to spot. When a house is wired to minimum code or less, there may not be enough receptacles, and the ones you have might not be in the right locations. If you have extension cords lying all over the place, you have a problem. The solution here is to add receptacles and circuits where you need them, not just where minimum code says to place them. If you don't know how, call a professional electrician to do the job.

Apart from the visual inspection, you need to verify whether the receptacles are wired correctly. Use a plug-in tester and test every outlet in the house. Note if the hot and neutral wires have been reversed and if the ground wires are connected. I've seen new houses with the ground wires cut off or not even tied to the receptacle. One inspector told me of a new house he checked where he just happened to measure the circuit voltage on one of the 120-volt receptacle outlets in the living room. It read 240 volts—anything plugged into that outlet would have gone up in smoke. Also be sure that you have ground-fault circuit interrupters (GFCIs) where they are now required: in the bathroom, along all kitchen countertops, islands, and peninsulas, and outdoors.

Damaged receptacles and switches

While you're checking the receptacles in the home, be on the lookout for any that are broken, as well as for switches that are broken. Remove the cover plates and check all of them. Any switches or receptacles with cracked plastic housings or bare metal showing should be replaced immediately.

Undersized wiring

A significant wiring problem is undersized wiring, where the wire gauge is too small for the loads being placed on it. Fourteen-gauge wire was—and is—fine for small loads, but it's not really meant to handle heavy appliance loads (more than 15 amps). For heavy loads, 12-gauge wire should be used.

I had one customer complain that the fuse would blow every time he plugged in his 1,500-watt portable heater. The circuit had 14-gauge wire and was protected with a 15-amp fuse. The heater was pulling 12 to 14 amps, which would have been fine if it were the only appliance on the circuit. But with the other loads being placed on the circuit, the load exceeded the current rating of the wire, causing the fuse to blow. The solution was to run a separate 12-gauge circuit for the heater. If you have a problem with breakers kicking off or with fuses blowing frequently, the cause could be undersized wiring.

In newer homes, a frequent mistake is tying 14-gauge wire into 12-gauge receptacle circuits. This is a common mistake made while wiring three-way switch circuits (for more on wiring switches, see Chapter 3). The only way to verify this problem is to pull the three-way switch out of the wall and see if 14 gauge is written on any cable insulation going into the switch box. If there is 14-gauge wire attached to the switch, and the power to the switch is controlled by a 20-amp breaker, you've got problems. Since 14-gauge wire cannot remain attached to a 20-amp breaker, the breaker must be changed to 15 amps. However, if the circuit needs the full 20-amp load, a new 12-gauge wire will have to be run, a job that may require a professional electrician.

Damaged wiring

Checking the physical condition of the wiring is an important part of the inspection. In general, wires should not be frayed, should not have insulation missing, and should not be burned or discolored in any areas. They should not be exposed in finished areas, where they could be damaged or where someone could touch them. And wires should not be bent around sharp corners or around sharp objects, such as ductwork.

Although a home's wiring is 99% hidden, there's one symptom that usually indicates a problem. If a breaker continues to kick off while no load is on the circuit, either a receptacle is bad or the wiring is. To find the source of the problem, first inspect the receptacles on the circuit as well as all splices. Make sure that the wires are secure in their terminals and that all splices are tight and not shorting against something. If they are loose, tighten them (make sure the power is off). If the screw terminals are broken, get a new receptacle.

Many times the problem can be attributed to the receptacle screw shorting against the side of a metal receptacle box. To prevent or remedy this, wrap a couple of turns of electrical tape around the sides of any receptacle going into a metal box. I sometimes even put a couple pieces of electrical tape into the back of the box to protect the wires as well.

If you don't find problems with the receptacles or splices, you'll have to check the cables. Look in the attic and crawlspaces for any cables pulled tight around sharp objects, such as ductwork, beams, and plates, or around tight corners. Bending wires tightly around corners can break the insulation or the wires, creating a loose connection that could overheat.

3

RECEPTACLES AND SWITCHES

Two of the most commonly miswired components of the house electrical system are receptacles and switches. I've made many service calls to rewire what someone has done wrong, and some of these calls still make me shake my head in amazement. It's really unbelievable the variety of ways a receptacle or switch can be wired incorrectly.

Often, a miswired switch or receptacle simply won't function. Two days ago I was called out to repair a switch that a homeowner just couldn't get to work. It took me almost two hours to straighten it out. He had a separate splice box that wasn't required, and the splice box was hidden in the wall, which is against code (splices are maintainable items and must be accessible). From there he brought in extra hot wires and neutrals that weren't needed and gave me one less wire than I needed at the light. At the light, he had all kinds of wires that had nothing to do with the switching circuit. In the end, I had eight wires left over that I didn't need.

In that situation, nobody's life was threatened. But a miswired switch or receptacle can also be dangerous. I went on one call in which the receptacle was so hot that it was glowing cherry red and on another call where the child safety inserts in one receptacle had melted. In both cases, the receptacles were cheap, residential grades with push-in terminals. The connections had worked loose, creating heat in the outlet and intermittent current flow. If the homeowners had not noticed these situations, a fire could have started.

I'm not trying to scare you so much that you hire an electrician every time you need a new switch or receptacle installed. Instead, I'm trying to encourage you to do these jobs yourself because installing a switch or receptacle is really not that difficult. It's just a matter of buying good-quality hardware and knowing how to run the wires and to hook them up properly. The focus of this chapter will be on standard 120-volt receptacles and switches. In it, you'll learn how to choose the right boxes for them, how and where to run the wiring for them, and how to wire and troubleshoot them.

The nonmetallic boxes on the left attach to studs via the integral nail holders or the bracket. Metal boxes like those on the right are available in a variety of sizes and volumes. The box on the far right has rounded corners so that it can be used in exposed areas. (Photo by Roe A. Osborn.)

CHOOSING ELECTRICAL BOXES

One of the first steps in installing a receptacle or switch is choosing the right box, and there is a wide variety to choose from. In general, both switches and receptacles use the same boxes (see the photo above).

Volume

The first consideration in choosing a box is the volume, or cubic inches of space inside the box. The National Electric Code (NEC) calls this space "cable fill" and specifies the amount of cable fill needed based on the size and number of wires entering and leaving the box and on the type of device the box will hold. If a box is too small, it will be overcrowded, resulting in broken or shorted wires and possible damage to the box's receptacle-holding threads.

When choosing a box, it's better to have too much volume instead of too little, which will give the wires plenty of breathing room. Having extra volume in the box also gives you the option of adding circuits in the future.

When picking a single box, you'll have a range between 16 cu. in. and 23 cu. in., with 18 cu. in. being most common. I recommend opting for the box with the largest-possible volume. For switching, you can choose double, triple, or quad boxes, depending on the number of switches to be installed at one location.

Nonmetallic or metal

When choosing a box, you also have to think about whether to buy nonmetallic or metal.

Nonmetallic boxes are made of PVC plastic, fiberglass, or thermoset and are now the most commonly used boxes in homes (they are my preference). Most single-gang (meaning it holds one device) nonmetallic boxes are made with integral nail holders and come with nails already in the holders. They install quickly and are less expensive than metal boxes (they also have their volumes stamped inside). An added bonus of nonmetallic boxes is that they are nonconductive.

Choosing Receptacles and Switches

Even if you can get a cheap receptacle installed without breaking it, the cheap brittle plastic will break sooner or later, as this one did at the top ground slot.

Many people don't realize that they have a choice about which type of receptacles and switches to buy. They just go to the mega-home store and pick a receptacle or switch out of a bin, without really taking a close look. But the quality of receptacles and switches varies, and choosing good-quality hardware is a must if you want it to last.

Quality is determined by the materials in the switch or receptacle and how they are put together. Since cheap receptacles and switches are made from brittle plastic, at least 10% of such will break apart as they are installed. The most common problems are having the bodies break and side screws stripping out in the body.

A good-quality receptacle will have a hard-to-break face, a wrap-around yoke for support, a heavy-duty body, a grounding clip, and heavier metal on the inside. When you're at the store, hold a light-duty receptacle in one hand and a heavy-duty one in the other. Just by feel, you can tell that the heavy-duty model is better.

A good-quality switch will have an unbreakable nylon toggle, a heavy-duty, corrosion-resistant brass- and nickel-plated yoke, large screw terminals on the side, and dual grounding options—a grounding terminal or a grounding clip.

The quality of receptacles and switches you will need depends on what kind of use they will be subjected to. For example, cheap residential-grade receptacles or switches are usually recommended by manufacturers for light-duty use, such as hallways or bedrooms. One problem with residential grade is that you get what you pay for. These devices don't last very long even with light use. As a matter of fact, they may not survive the installation—I've had countless numbers of cheap pieces of hardware break apart in my hands before I could even get them in the wall. And many of my service calls involve replacing residential-grade receptacles and switches that fell apart or wore out quickly.

Another problem with residential-grade receptacles is that many of them are equipped with push-in terminals to connect the wires. This type of connection is for lazy electricians and do-it-yourselfers who are in a hurry and simply push in the wire to the terminal. The mechanical connection inside a push-in terminal is not very good, and over time the connection may become loose, and eventually the wire may pull out completely. I had one service call where the homeowners thought their house was haunted because of the problems these receptacle connections gave. If you have a receptacle with this type of connection mechanism, don't use it. Attach the wires to the screw terminals.

For high-abuse areas, such as kitchens or garages, heavy-duty hardware—called spec or commercial grade—is recommended. But beware! Just because a re-

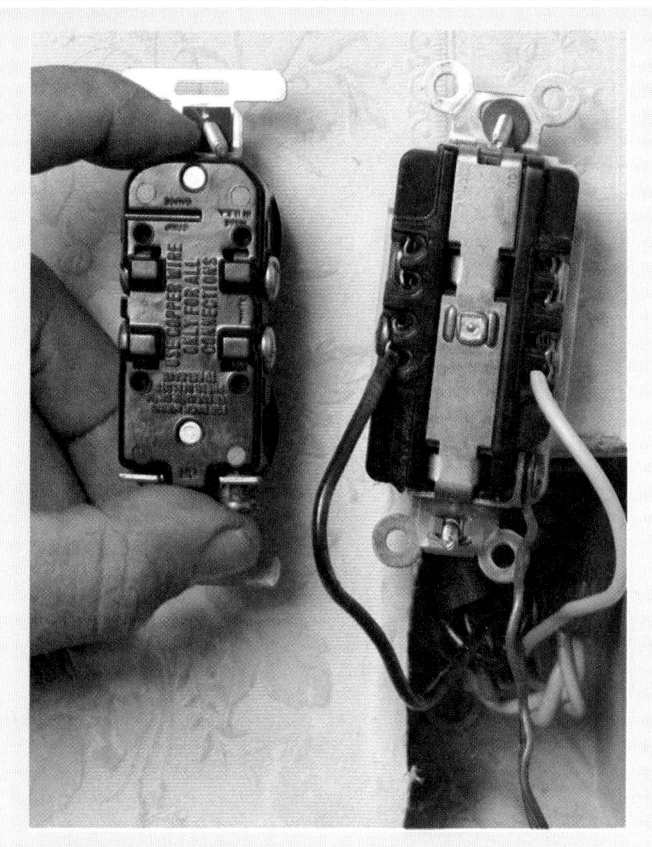

The commercial-grade receptacle on the right has a heavy-duty body and a wraparound yoke and provides eight connections. The residential-grade receptacle on the left is made of cheap plastic and provides only four connections.

A good-quality receptacle will have a hard-to-break face and a grounding clip (bottom), which provides an automatic ground for the receptacle if it is installed in a grounded metal box.

ceptacle or switch is commercial grade doesn't mean it is made of nylon or other hard-to-break material. I put expensive commercial-grade receptacles in my garage when I built my house. They all broke around the grounding slot, and I had to replace them all.

I not only recommend commercial grade for high-abuse areas, but I also recommend it for use throughout the home. Unfortunately, buying commercial-grade hardware can be expensive ($2 and up apiece), depending on the number of switches and receptacles you need. But I think buying a better receptacle or switch is worth the money because you'll have fewer problems down the road, meaning you won't have to call a guy like me to come out and repair or replace what you've installed.

An adjustable box can be moved in or out with the turn of a screw, which makes it easy to achieve a perfect, flush fit with the finished wall. (Photo by Roe A. Osborn.)

The problem with nonmetallic boxes is that they are not as durable as metal ones. That means they must be installed inside walls, not in exposed areas where they could get bumped and broken.

Metal boxes used to be the standard of the industry, but no more, although they still offer a few distinct advantages. First, they are stronger and offer a wider variety of styles and sizes. They should be used in exposed areas because they'll be able to stand up to bumps and bruises (you can even get boxes with rounded edges for surface mounting in garages and basements). But metal boxes are conductive, so they need to be grounded by a pigtail off the ground-splice connection. They also pass fire and heat quicker than a nonmetallic box, especially with the knockouts removed. I use metal boxes only when they are necessary. And if I must use a metal box, I line the sides with electrical tape—sometimes even the back—so that I don't accidentally short out any hot terminals against the box.

Specialty boxes

Installing boxes can be difficult. You have to be sure that the box sticks out from the stud so that it will be flush with the finished wall. You also have to be sure not to distort or break the box while nailing, all the while keeping the front of the box parallel to the wall.

To make installation easier, manufacturers designed a nonmetallic box with a bracket that automatically sets the box parallel and the correct distance from the stud. You can also buy an adjustable box, which, after it is nailed to the stud, can be moved in or out with the turn of a screw to fit flush to the finished wall (see the photo at left).

For remodeling work, manufacturers offer cut-in boxes (also called old-work boxes), which allow you to install a box without tearing up the wall to find a stud to nail to. A cut-in box is inserted into a hole cut into the wall and holds itself in place by sandwiching the wallboard between a bracket on the back and the drywall ears on front (see the photo on the facing page). The hole in the wall must be cut exactly the same size as the box, though, or the box will fall through.

Installation

Receptacles and switches should be installed for convenience. There is no general code requirement. Unless I'm told otherwise, for new installations I mount receptacle boxes one hammer height from the bottom plate (see the drawing on the facing page), or about 12 in. to

A cut-in box holds itself in place by sandwiching the wallboard between the brackets on back and drywall ears on front. (Photo by Roe A. Osborn.)

Finding the Receptacle-Box Height

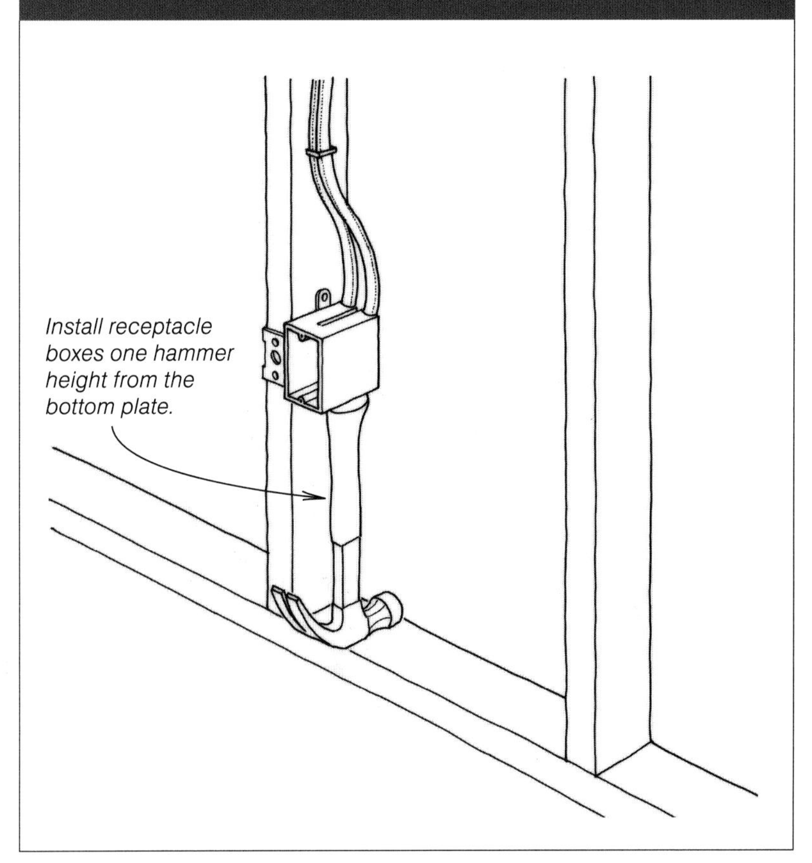

14 in. from the floor. For handicapped access, consult your local codes. Receptacles and switches for dedicated circuits require special treatment (see Chapter 9).

Switches must be no higher than 6 ft. 7 in. from the floor, according to the NEC. I think a good spot to install switch boxes is 4 ft. off the floor. For receptacles and switches located above countertops, make sure the box is at least 1 in. above the top of the backsplash.

As I said before, the front of the box should be flush and parallel to the finished wall, which could be frustrating if you are installing a box with integral nail holders. To make life easier, use a nonmetallic box with a nail bracket or an adjustable box. Also, be careful when driving the nails—one missed hammer blow can destroy a nonmetallic box.

Receptacle and switch boxes have small holes at the top and bottom to accept the hold-down screws of the receptacle or switch. They also have four knockouts on them that can be popped out to accept incoming or outgoing cables.

Which Cable Is Which?

If you always bring power into a box via one particular knockout and run the outgoing cables through the others, you'll be able to verify at a glance whether the box has a hot cable or not.

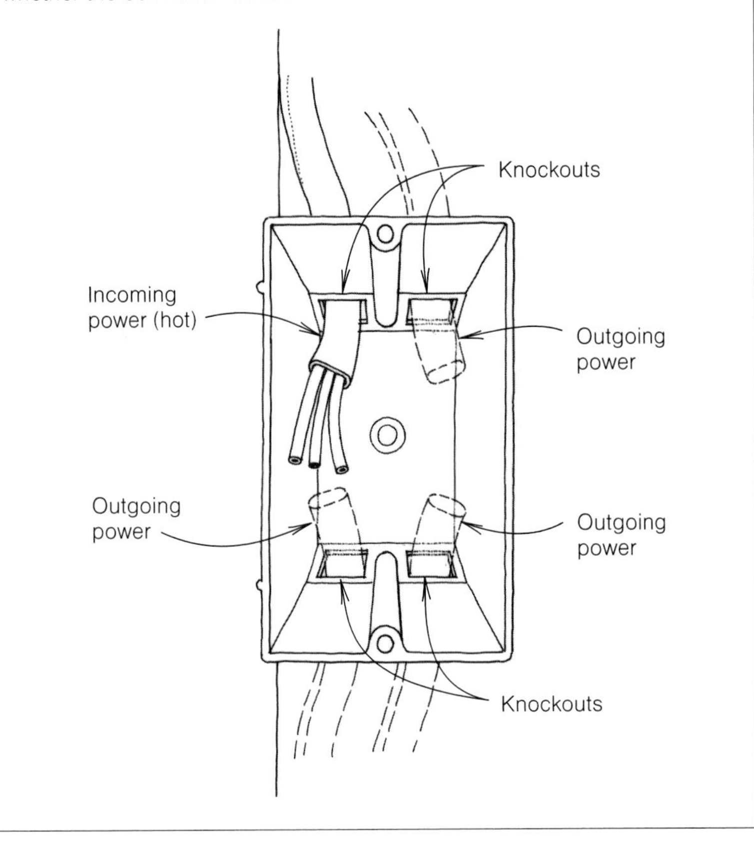

One problem I used to have when I first started out was remembering which cable was bringing in power to the box and which ones were delivering it to other locations. Here's a tip to help you avoid this kind of forgetfulness. Always bring in power to the box through one particular knockout (I use the one on the upper left). This way you will always be able to verify at a glance whether the box has a hot cable or not (see the drawing above).

WHAT'S BEHIND THE WALLS

House wiring is hidden behind the finished walls, so it's important to know the routes if you're going to do any electrical work (it also pays to know the routes if you plan on doing any remodeling so that you don't cut any power cables by accident).

The wiring routes are pretty standard. From the main panel, cables run through the attic, the basement, or the crawlspace, entering these areas through holes cut through the top or bottom plates of the walls. Then they run vertically along studs and horizontally through them to power receptacles or switches (or light fixtures and appliances). Cables are stapled to the center of studs every 2 ft. or 3 ft. so that nails and screws driven into the wall will not harm them.

Cables entering or leaving boxes are required by code to be stapled to the stud within 12 in. of a metal box and within 8 in. of a nonmetallic box. Holes for cables running horizontally through walls should be drilled high enough to allow the cable to be stapled to the stud above or below boxes. Cables also run over or under doors and windows (see the drawings on the facing page).

Running cable from the main panel can be difficult, time-consuming work if you've never done it before or don't have the right tools for the job. One way to save yourself headaches—and to prevent damage to cables and the house structure—is to hire a professional electrician to pull the cable to its destination. Then you can save some money by installing the switch or receptacle and making the connec-

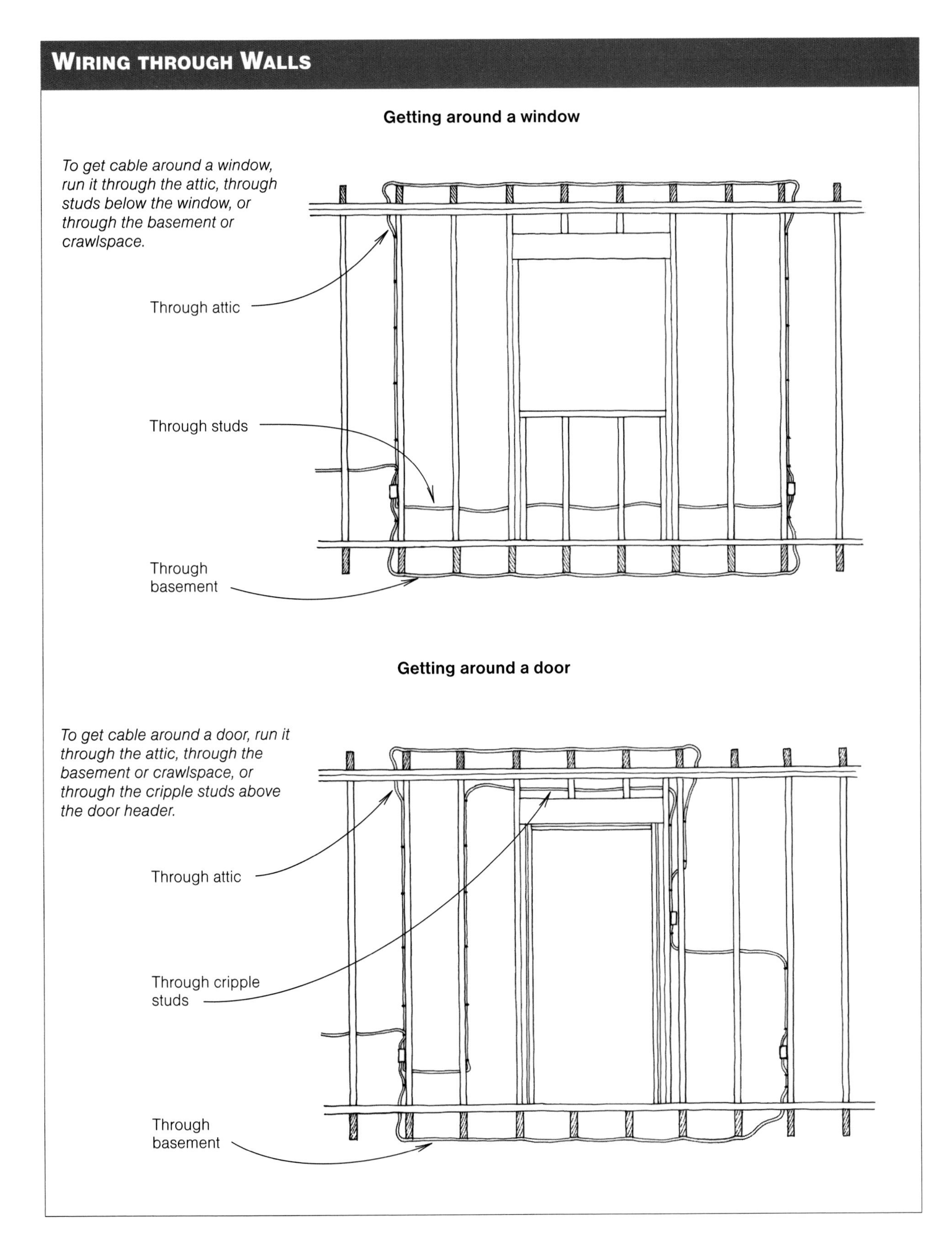
WIRING THROUGH WALLS
Getting around a window
To get cable around a window, run it through the attic, through studs below the window, or through the basement or crawlspace.
Through attic
Through studs
Through basement
Getting around a door
To get cable around a door, run it through the attic, through the basement or crawlspace, or through the cripple studs above the door header.
Through attic
Through cripple studs
Through basement

Running Cables through Existing Walls

Installing switches and receptacles in walls that are already finished can be tricky. But here are a few tips that can reduce the number of holes you cut into the walls and ceiling.

When installing a new receptacle or switch, the easiest way to run the cables from point to point is at the base of the wall (see the drawing below). Say you're installing a new receptacle that will be wired from another in the same room. First, draw a line along the top of the baseboard molding and then remove the molding. If paint is holding the molding to the wall, use a utility knife to break the seal.

After the molding has been removed, turn off the power and cut away the wallboard, staying beneath the pencil line. Install the new box, drill holes through the studs, and pull cable from the existing receptacle to the new one (leave about 6 in. of slack in each box). Connect the old receptacle to the new one, then replace the cut-away wallboard, and reinstall the molding.

If you need to wire around an existing door but don't want to cut away all the wallboard above the header and can't get through the attic or basement, run the cable through the shim space around the door (see the top drawing on the facing page). Remove the door trim and the baseboard molding. Turn off the power and pull the cable through the studs and the shim space. You'll have to chisel the shims enough to fit the cable and then protect the cable from nails and screws with $1/16$-in.-thick metal plates. After running the cable, reinstall any wallboard or trim you removed.

A similar system can be used if you are installing a new light and switch in a room and are powering them off an existing receptacle (see the bottom drawing on the facing page). First, turn off the power. Then remove the baseboard molding and cut away the wallboard to below the pencil line. Install the new switch box and the ceiling-fixture box. Cut away a section of wallboard at the top of the wall so you can drill through the double top plate (if you have an accessible attic, you won't need to do this). Also cut away enough of the ceiling to be able to pull the cable up through the plate. Drill holes in the studs and in the plate and run the cable from the existing receptacle to the switch box, then run another from the switch box to the light fixture. Make the connections, and replace any sections of wall, ceiling, or molding you removed.

Routing cable behind baseboard trim

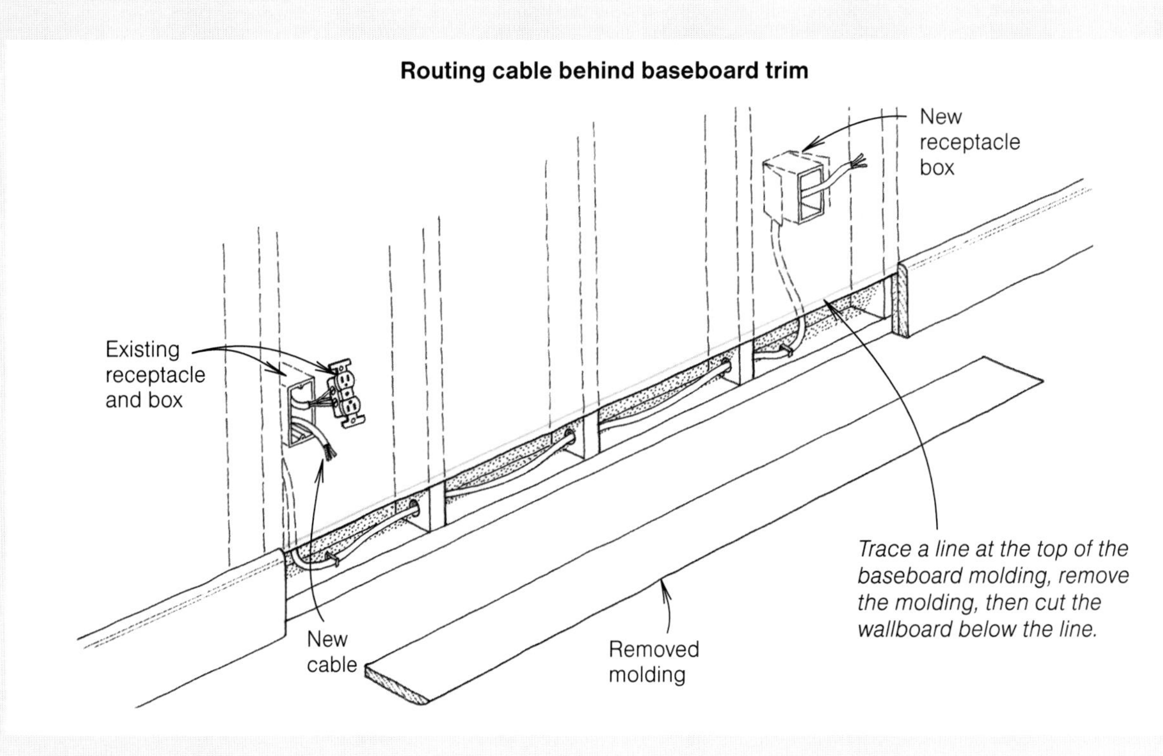

Routing cable around a door

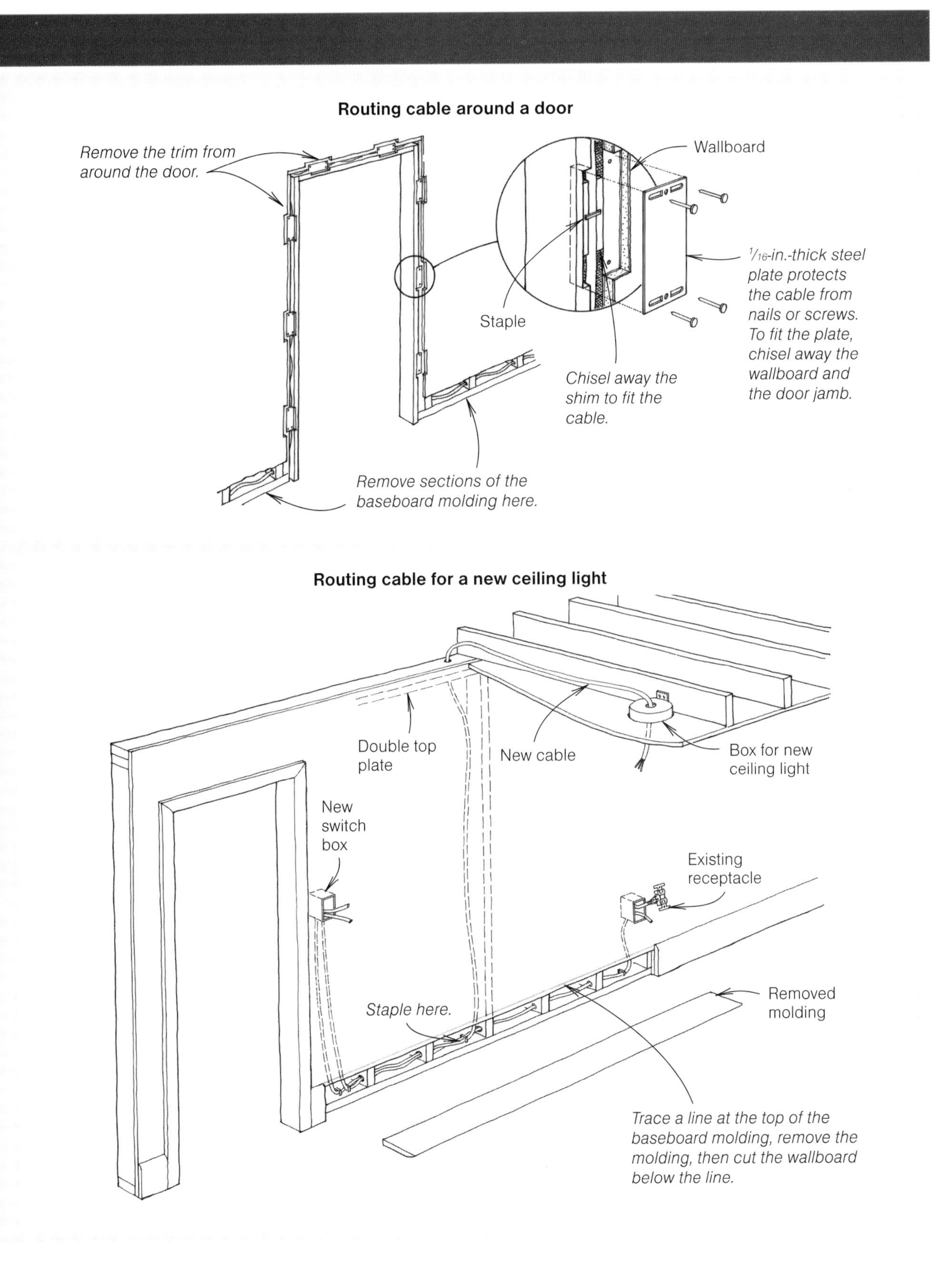

tions yourself. Now let's see how it's done. Before beginning any electrical work, remember to remove the power.

WIRING RECEPTACLES

The purpose of a receptacle is to provide access, or an outlet, to electrical power. Receptacles come in 15- and 20-amp configurations (see the photo below).

Most homes have 15-amp receptacles in them, even if they are wired into a 20-amp circuit. This is because it is assumed that each receptacle will be used at something considerably less than 15 amps. A 15-amp receptacle has two vertical slots and the grounding slot. A 20-amp receptacle is for appliances that draw a lot of current, such as some air conditioners and refrigerators. It has a horizontal slot on the wide, neutral slot to accommodate the special plugs on these appliances. Install a 20-amp receptacle only on a circuit with 12-gauge cable and protected by a 20-amp circuit breaker or fuse. Installation procedures for both 15- and 20-amp receptacles are the same.

The 20-amp receptacle on the left has a horizontal slot in the neutral slot that accommodates the special plug of an appliance that draws a lot of current. The other receptacle is a standard 15-amp unit. (Photo by Roe A. Osborn.)

Wire stripping

The first step in any installation is to bring the cable to the box. Pull it enough that you have about 6 in. of slack hanging out of the box. This will give you enough room to strip the cable end and expose the wires inside.

A utility knife with a sharp blade works well for cutting open the insulation on the cable. Start about ¼ in. from where the cable enters the box, and slice it down the middle of the flat side (see the left photo on the facing page). Be careful not to damage any wires inside the cable. Peel back the insulation and the heavy cardboard in the cable and slice it away as close as possible to where the slit began. I have repaired many jobs where the installer had a tough time getting the receptacle into the box, even when there was plenty of volume. The installer had not stripped the insulation away from cable back to within ¼ in. of where the wires entered the box. Don't make the same mistake.

Then cut all like wires to the same length and remove about ½ in. of insulation from any insulated conductors (see the right photo on the facing page). This can be done with a utility knife, but wire strippers work better and don't damage the conductor inside the insulation. With the wires stripped, you're ready to begin connecting them to the receptacle.

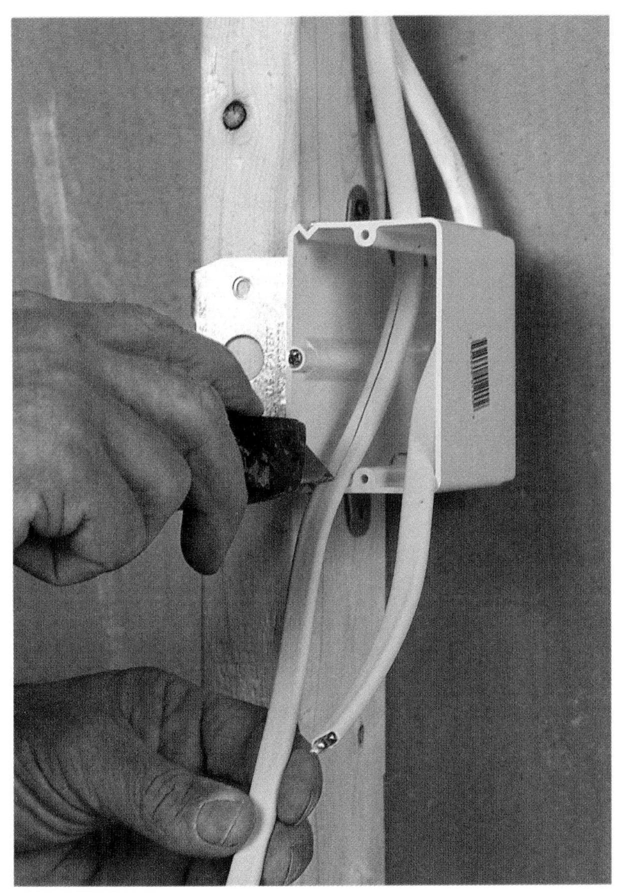

Cut the cable sheathing along the flat side with a utility knife, being careful not to damage the wires inside. (Photo by Roe A. Osborn.)

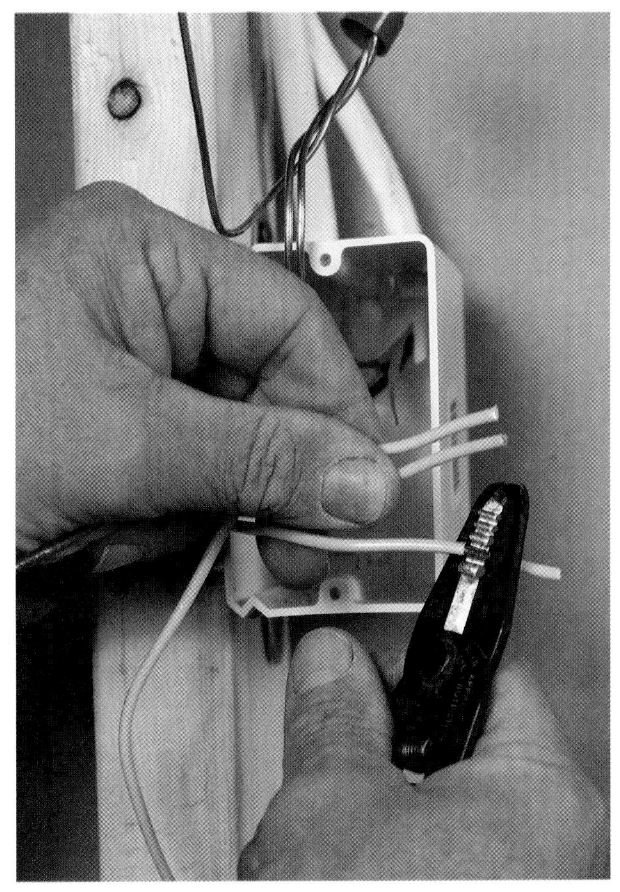

Strip about ½ in. from the end of any insulated wire in the cable using a knife or wire strippers. (Photo by Roe A. Osborn.)

Single receptacle

Power is accessed through the slots on the front of the receptacle. A receptacle has a narrow slot, a wide slot, and a grounding slot. If you are wiring just a single receptacle, simply attach the colored wires to their appropriate terminals.

The narrow slot is internally wired to the brass-colored screw terminals on the side of the receptacle. The black hot wire, which brings power to the receptacle from the main panel, is to be secured to one of these terminals (see the left photo on p. 38). To secure it, first twist the end of the wire in a small clockwise loop using needle-nose pliers. Back the screw out of the receptacle and slip the loop around the screw shaft so that the open section of the loop is on the right-hand side. Make sure most of the loop is under the screw, then tighten it down. The counterclockwise turning of the screw will grab the loop and secure it tightly. When tightening the screws, be careful not to strip the terminal's threads. If you do, throw away the receptacle; it is no longer good.

The wide slot on the receptacle front is internally connected to the silver-colored screw terminals on the opposite side of

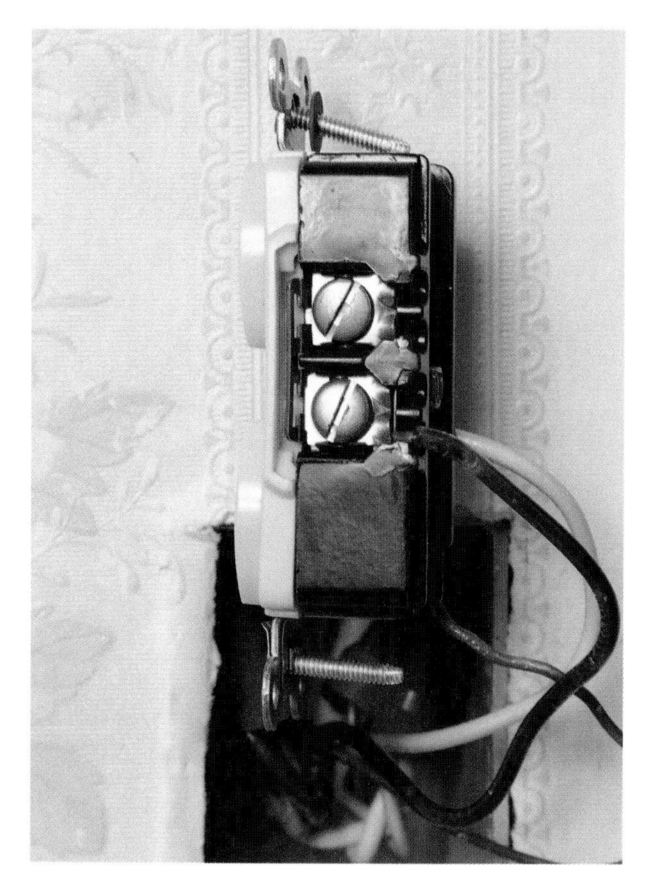

The black hot wire connects to one of the brass-colored screw terminals, which connects to the narrow slot on the front of the receptacle.

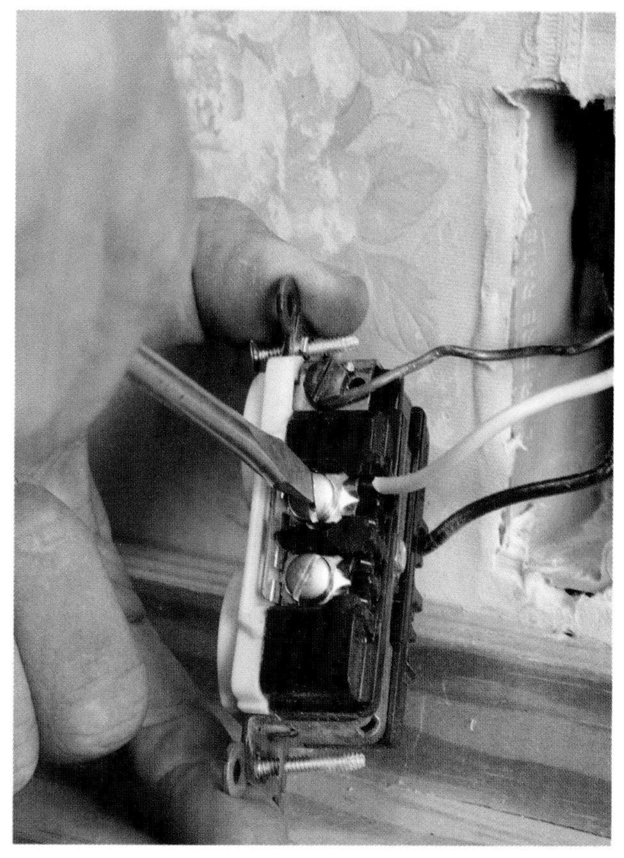

The white neutral wire connects to one of the silver-colored screw terminals, which connects to the wide slot on the front of the receptacle.

the brass screw terminals. The white neutral wire provides a return path (to the main panel) for the electricity provided by the black wire. Connect it to one of the silver terminals using the same attachment method as before (see the right photo above).

The grounding slot on the front of the receptacle connects to the green grounding screw on the back of the receptacle. The bare copper wire, called the equipment grounding conductor (more commonly called the ground wire), connects here. You'd be surprised how many people don't know that simple fact. I made one service call to troubleshoot a new renovation where the previous installer had cut off every ground wire where it entered the box. Unbelievable. (I know an inspector who sees this at least once a month.)

Receptacle strings

If you are wiring more than one receptacle in a circuit (called a string), there are two ways to do it: in series or in parallel.

Wiring receptacles in series is the most common method of wiring a string of receptacles on one circuit (see the top drawing on the facing page). In a series circuit, incoming current flows through all the receptacles of the string. The advantage of this string is that it's easy to hook up. The disadvantage of this wiring

Receptacle String in Series

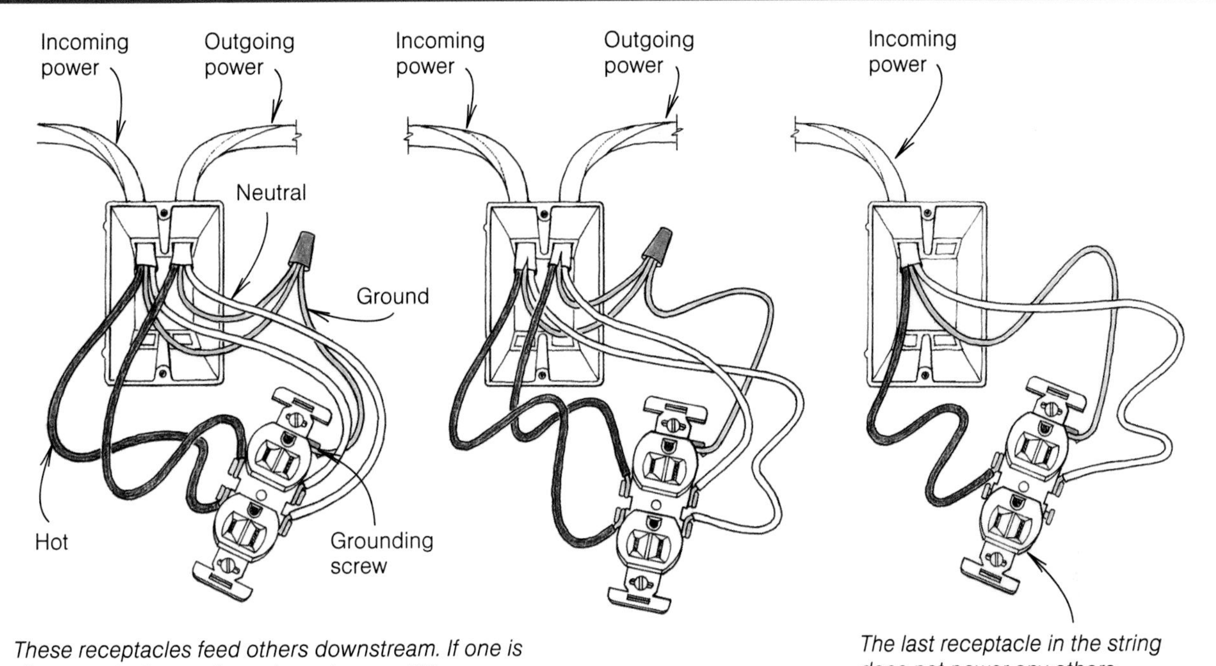

These receptacles feed others downstream. If one is disconnected, any others downstream will lose power.

The last receptacle in the string does not power any others.

Receptacle String in Parallel

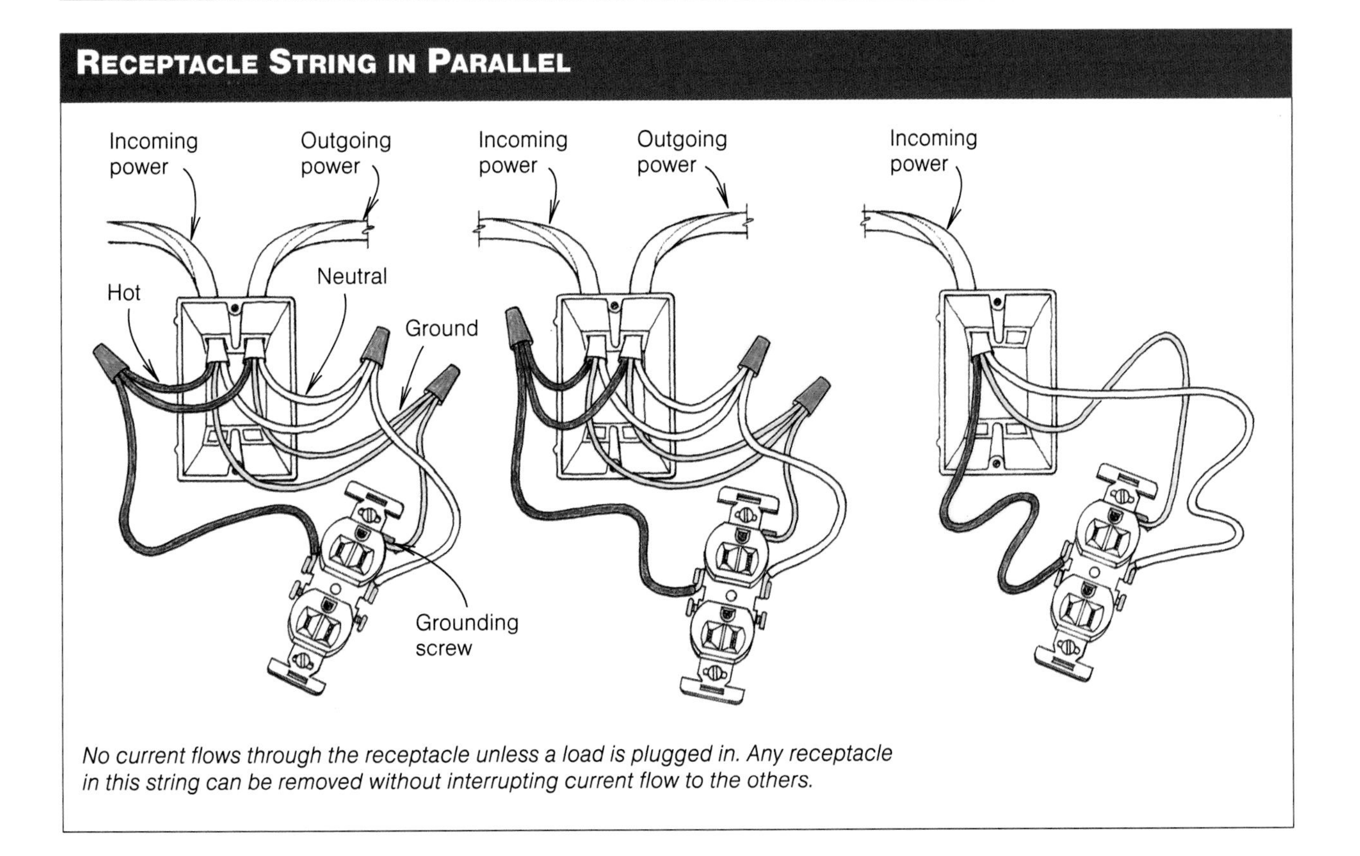

No current flows through the receptacle unless a load is plugged in. Any receptacle in this string can be removed without interrupting current flow to the others.

Splicing with a Wire Nut

Knowing how to make a good electrical splice is a must for anyone working with electricity. A loose connection can cause intermittent current flow and can generate enough heat to melt wire insulation. The most sound method is to splice using a wire nut (see the Glossary on p. 146), which provides a strong mechanical connection. Here's how to splice with a wire nut.

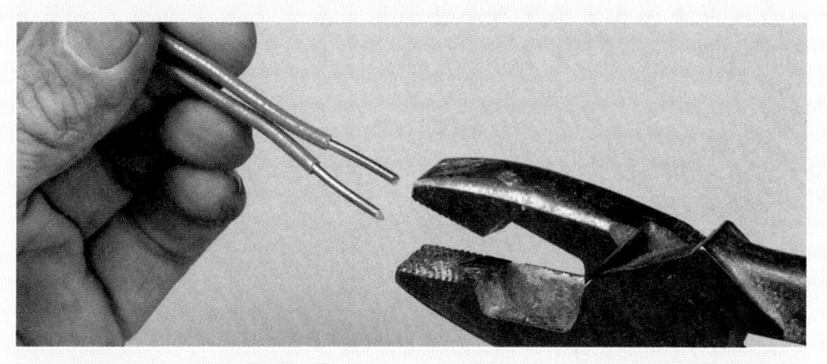

Cut all wires to be spliced the exact same length and strip off around ½ in. of insulation. Then twist the wires together in a clockwise motion using pliers.

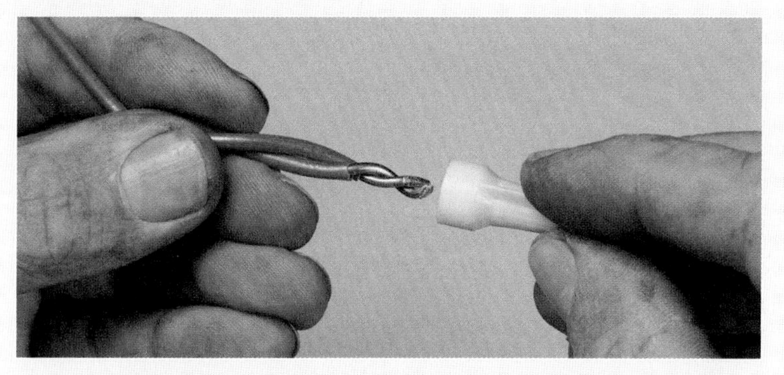

Screw a wire nut on the twisted wire end—also in a clockwise motion. Twist the wire nut until firm pressure is obtained. If the wire nut just spins as you turn it, without catching, either the wire nut is too big or a wire end is bent.

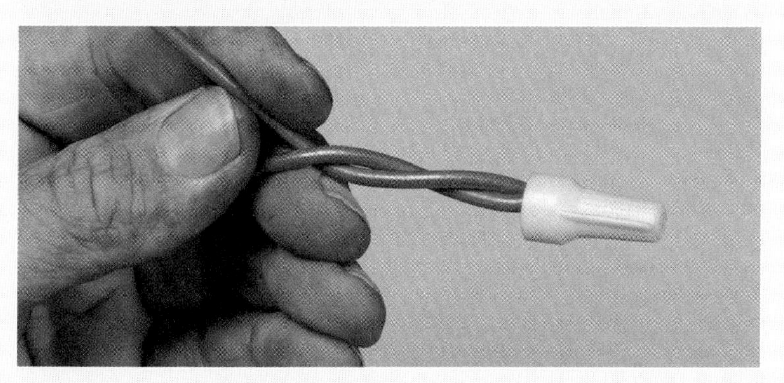

No bare wire should show below the skirt of the wire nut.

setup is that if a wire comes off one receptacle, power will be cut off to that receptacle and every other receptacle downstream (those powered off it).

A much better method of wiring receptacle strings is to wire them in parallel (see the bottom drawing on p. 39). In a parallel string, no current flows to the receptacle unless a load is plugged into it. Any receptacle can be removed without cutting off power to those downstream. Wiring a parallel string takes a bit more work than wiring a series string.

The first step in a parallel string is to cut 6-in. pieces of white, black, and ground wire and strip ½ in. of insulation at each end. Then splice the short length of black wire (called a "pigtail") to the incoming black wire and the outgoing black wire using a wire nut. Do the same for the other pigtails and wires: white to white, ground to ground (for more on splicing, see the sidebar at left).

Loop the end of each pigtail and attach it to the appropriate terminal: black to the brass terminal, white to the silver terminal, and the bare ground to the green grounding screw.

Pushing the receptacle back into the box

After hooking up the receptacle, push it carefully into the box. If you've installed a big enough box and if the splices have been pushed neatly into the back of it, the receptacle should slide in easily. It helps to bend the pigtails into a Z shape so that they can be folded accordion style into the box.

If you are having trouble pushing the wires back in, it's probably because the box is too small. Don't force the issue. I was on a job once when one of my helpers tried to stuff two three-wire cables and a ground-fault circuit interrupter (GFCI) receptacle into a small metal utility box. Sparks flew, one wire totally burned, the breaker tripped, and a $20 GFCI bit the dust. Don't make the same mistake.

After pushing the receptacle into the box, restore power and test it. If the receptacle is working properly, put on the cover plate.

Wiring a GFCI receptacle

A GFCI receptacle protects people against receiving deadly shocks in the event of a ground fault (see the Glossary on p. 146). Often when a ground fault occurs, the current takes a path to ground that has the least resistance—usually a human. Simply put, a GFCI receptacle can sense when a ground fault has occurred on a circuit and opens, or interrupts, the circuit faster than a heartbeat—quick enough to save a life.

In new construction, GFCI receptacles are required in areas that are damp or that have plumbing, such as kitchens, bathrooms, crawlspaces, and even outdoors. Older homes may not have any kind of GFCI protection in these areas, but they should. GFCI receptacles can also be used to protect ungrounded two-prong outlets.

All GFCI receptacles have two screw terminals marked as line (input) and two more screws marked as load (output), as shown in the photo above. The incoming black (hot) and white (neutral) wires connect to the two line terminals. The two load terminals are used only if you want downstream receptacles to have GFCI protection as well. Never mix these wires because the GFCI may not protect you properly. The ground terminal is identical to a typical receptacle ground terminal and has nothing to do with life protection.

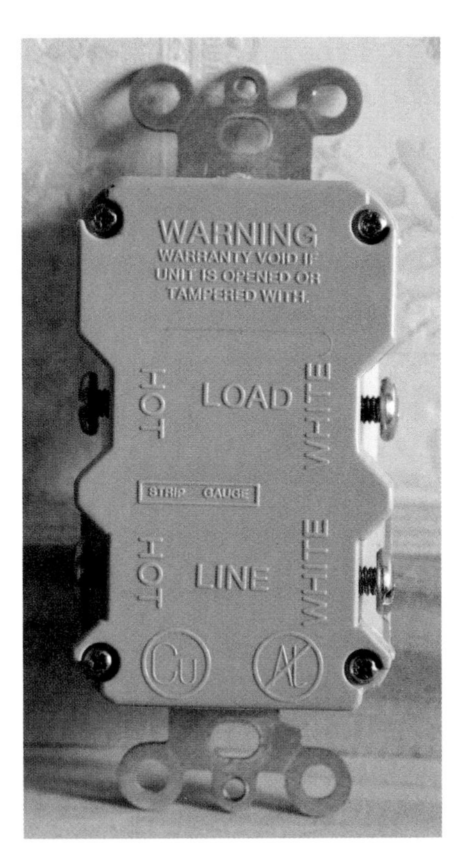

The line terminals on a GFCI receptacle are for incoming power. The load terminals are for outgoing protected power.

GFCI receptacles have a test and reset button and a sticker that reminds you to test it monthly. You can only test a GFCI receptacle while it is hooked up—you can't take a unit off the store shelf or out of its box and press the buttons to see if it works. The test button puts a simulated ground fault on the neutral to verify that the unit will work. If it's working properly, the reset button will pop out, and power will be removed from the circuit. Press the reset button back in, and power will be restored.

As with standard receptacles, GFCIs are available in residential and commercial grades. Residential grades don't last very long. I installed some in my house a couple years ago, and they all are dead now. What I should have installed then and do install now is commercial grade (also

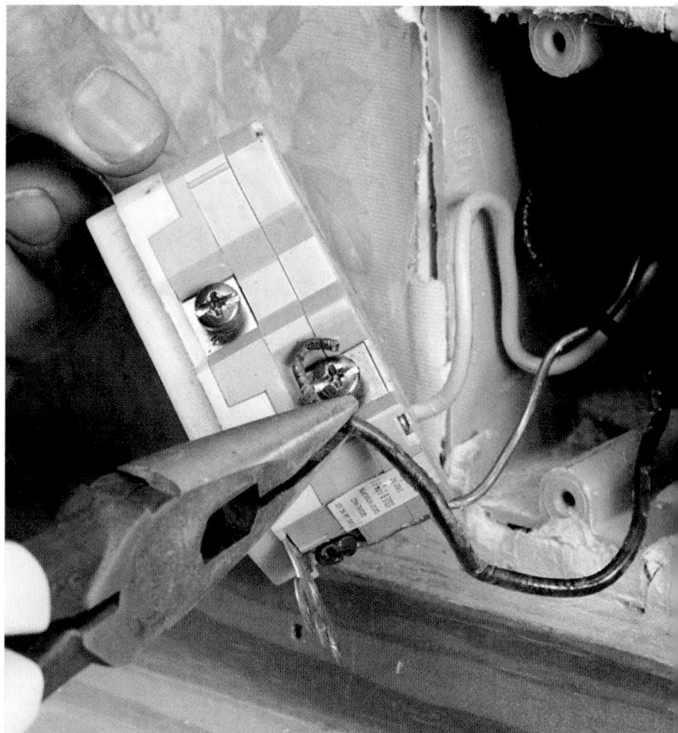

Attach the white neutral to the white terminal on the line side of the GFCI (left) and attach the ground wire to the grounding terminal. Then attach the black wire to the hot line terminal (right).

called industrial grade). Live and learn. You can get commercial units at wholesale electrical stores for less than $20.

Installation of a GFCI receptacle is similar to the process described previously. To replace an old receptacle with a GFCI receptacle, turn off all power, and remove the receptacle from the wiring.

If the receptacle has only one black and one white wire (plus ground) attached to it, you can make a direct transfer of these wires to the GFCI receptacle (see the photos above). First connect the white wire to its terminal on the line side. Then connect the black wire to the hot terminal on the line side. If there are two black and two white wires attached to the existing receptacle, splice like colors together (black to black, white to white, and all grounds) and run a pigtail from each splice to the line terminals of the GFCI and the ground terminal. Again, you can't have more than one wire under any terminal. If you're putting in a new GFCI receptacle, install the box, run the cable, and connect the wiring as I just described. After installation, restore power and test the GFCI receptacle to see if it works.

WIRING SWITCHES

The purpose of a switch is to connect power to a load, normally a light, when it is thrown to the on position. If you are installing a new switch, put in the appropriate box, run the cable, strip it, and then strip the ends of the insulated wires inside (see the right photo on p. 37). The three most common switches used in the home are single-pole switches, three-way switches, and dimmer switches.

Single-pole switch

A single-pole switch is the most common one used in a residence. It has a grounding terminal on the bottom and two screw terminals on its side: one for the wire carrying the incoming power, and the other for the wire that carries switched power to the load. It makes no difference which wire connects to which terminal. When the switch lever is raised, the wires are electrically connected together within the switch.

There are three methods of wiring a single-pole switch. The simplest method has the power being brought to the load, then to the switch box. The wires are spliced at the load to wires coming from the switch (see the left drawing on p. 44). The thing to watch out for here is that the white wire can be hot because it carries current from the load splice to the switch. To indicate that it is a hot wire, tape the white wire with black tape at the splice and at the switch box. This is not required by code—it's just common sense.

Another wiring method has incoming power going to the switch box (see the right drawing on p. 44). I prefer this method because it allows easy access to the power cable. The incoming hot wire connects to the switch, as does the black wire from the outgoing cable, which carries switched power to the load. In this case, the white wires are spliced together and remain neutral, so no black tape is required. Also, the ground wires are spliced together, with a pigtail to the switch.

Life-Saving Switch

Imagine an ambulance is on its way to your house, and only seconds stand between life and death. The ambulance turns onto your street, but it's dark, and the driver can't read the house numbers. Precious time is lost trying to find the right house. This situation can be avoided by installing a new specialty switch that can be turned to a position that makes the outside lights flash. The flashing lights make it easy for an emergency crew or police to distinguish your house from all the rest on the block.

To install, all you have to do is replace your existing outside light switch with the flasher switch. Once power is off, remove the existing switch from the outlet box. Remove the two wires attached to the switch and splice them with wire nuts onto the flasher switch's two wires. That's all there is to it. Now put the switch back into the box and reinstall the cover plate. The switch will still have a down off position and an up on position. However, when turned halfway between the two, the lights will start to flash.

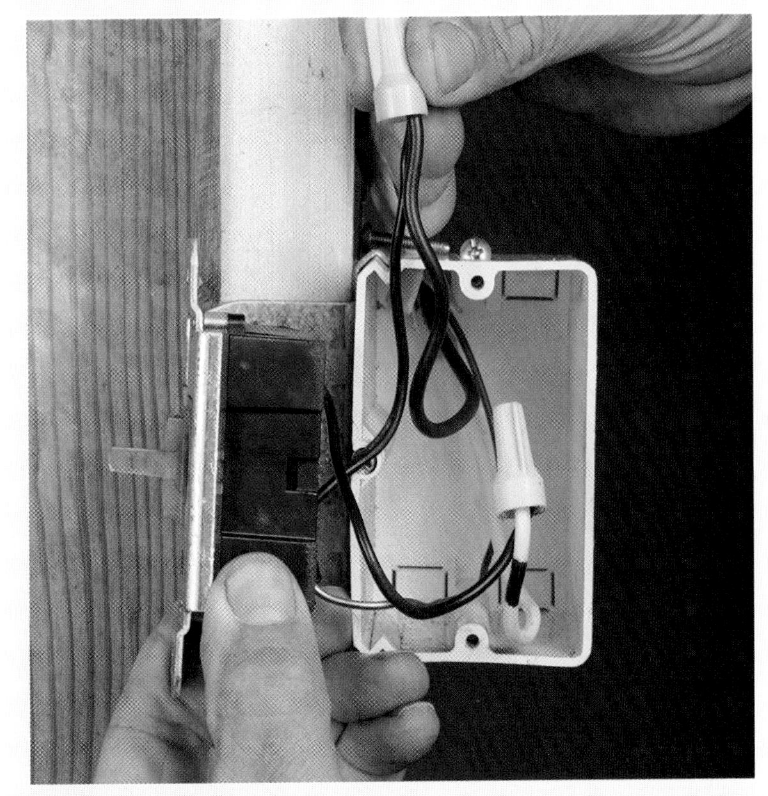

To install this life-saving switch, turn off the power, remove the wires from the existing switch, and splice them to the flasher switch's pigtails. You don't have to worry about polarity.

Three Methods of Wiring a Single-Pole Switch

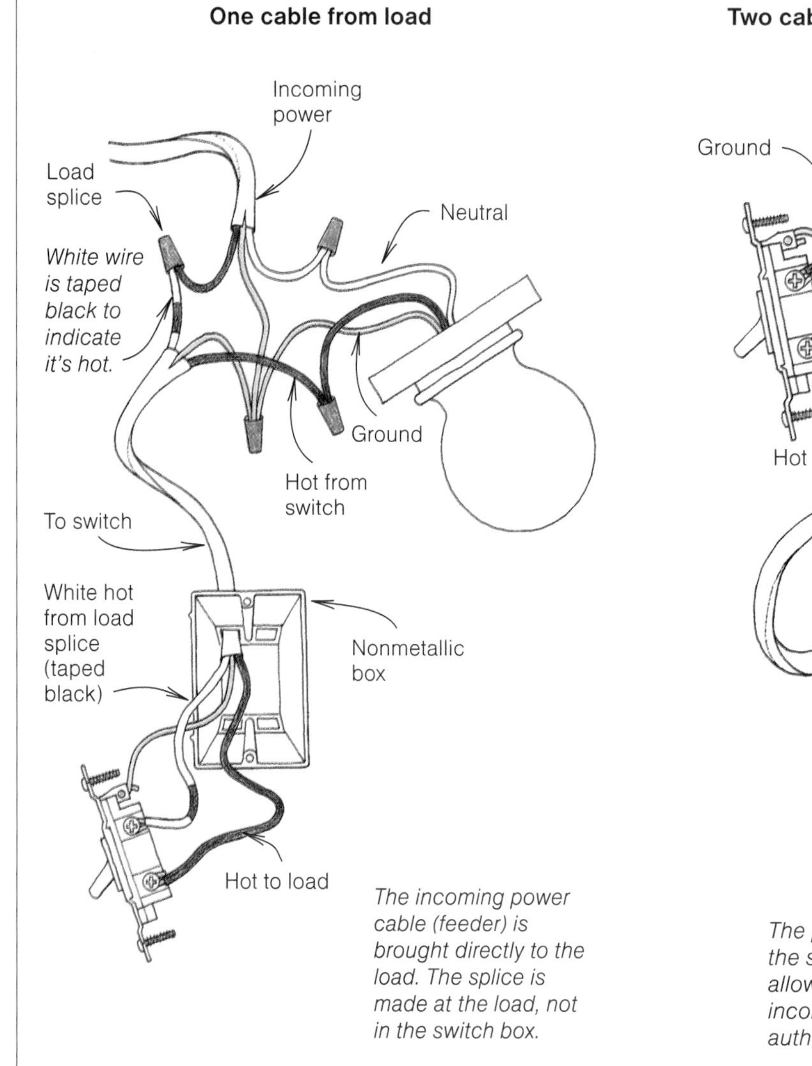

The incoming power cable (feeder) is brought directly to the load. The splice is made at the load, not in the switch box.

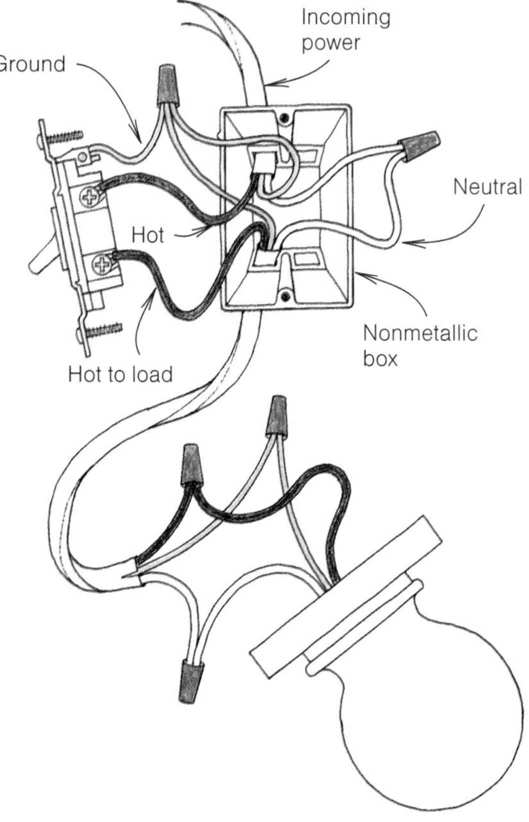

The power is brought through the switch to the load, which allows easy access to the incoming power. This is the author's preferred method.

If you want to wire a receptacle from the switch box, you must use my preferred wiring method, which brings power directly to the box. Run a cable to the switch box from the receptacle location. If you don't want power to the receptacle to be controlled by the switch, wire it as shown in the drawing on the facing page. In this situation, the switch will still control power to the load, but the receptacle will have constant power.

Three-way switch

The purpose of a three-way switch is to provide load control, normally a light, at two locations. The back of the switch has three terminals: two are called travelers and a third, a dark-colored screw, is called the common, or COM terminal. The COM terminal is like a tongue: It laps, or connects, alternately between the other two screw terminals.

Three cables: incoming power, load, outgoing power

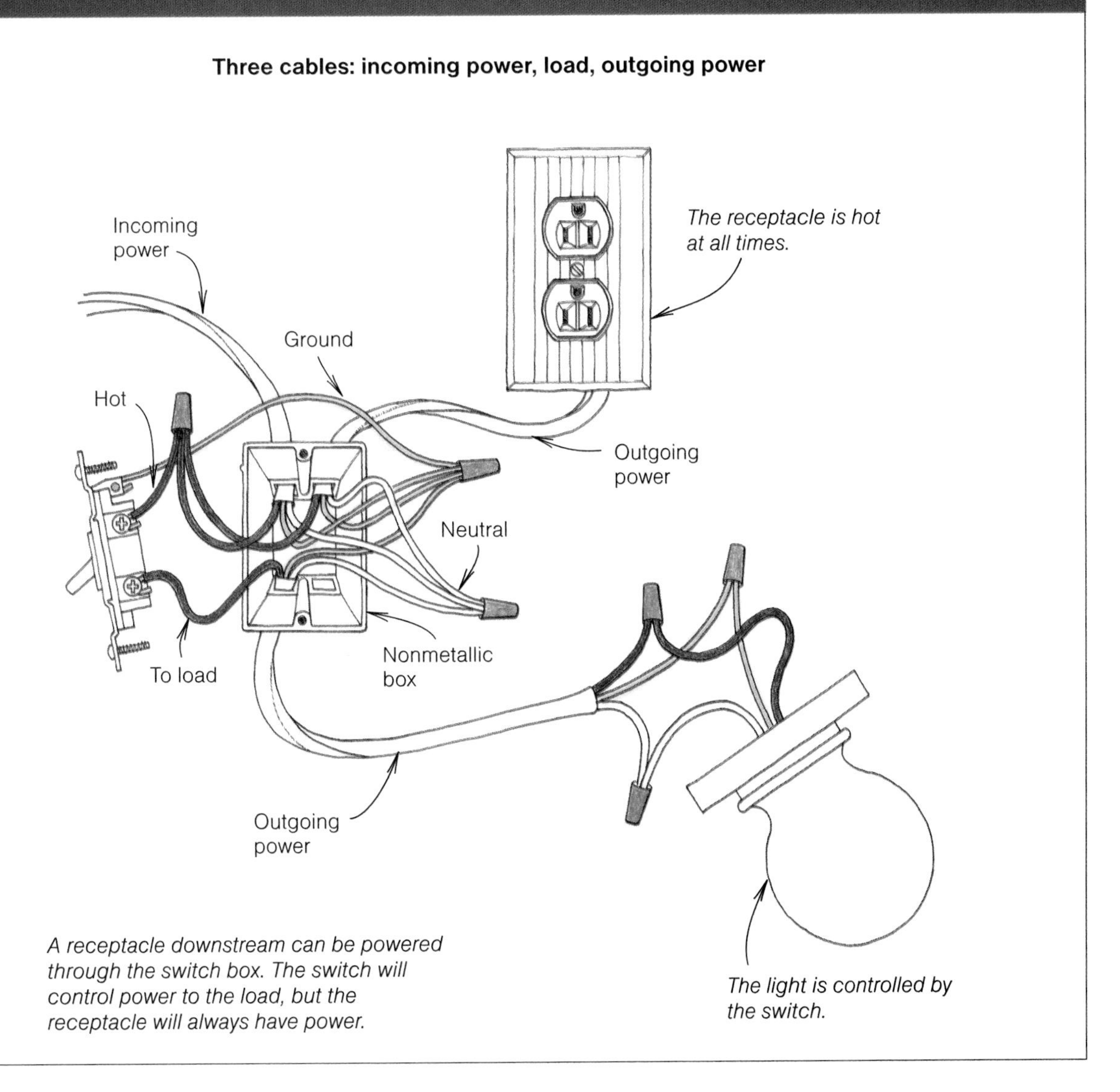

A receptacle downstream can be powered through the switch box. The switch will control power to the load, but the receptacle will always have power.

Things can get confusing when installing a three-way switch, but when all else fails, remember my three basic rules of three-way switch wiring. First, the hot black wire always goes directly to a COM terminal on one of the switches. Second, the neutral of the incoming power always goes directly to the load neutral, with the other load wire (black) going directly to the COM terminal of the other three-way switch. Third, the two unused traveler terminals of the first three-way switch connect directly to the two unused traveler terminals of the second three-way switch (see the drawing on p. 46). It doesn't make any difference which wire goes to which of the two terminals—there is no polarity (see the Glossary on p. 146). In this instance, one of the white wires is a traveler and will be hot, so it's a good idea to tape it black so that a person working on it will know it's hot.

Wiring a Three-Way Switch

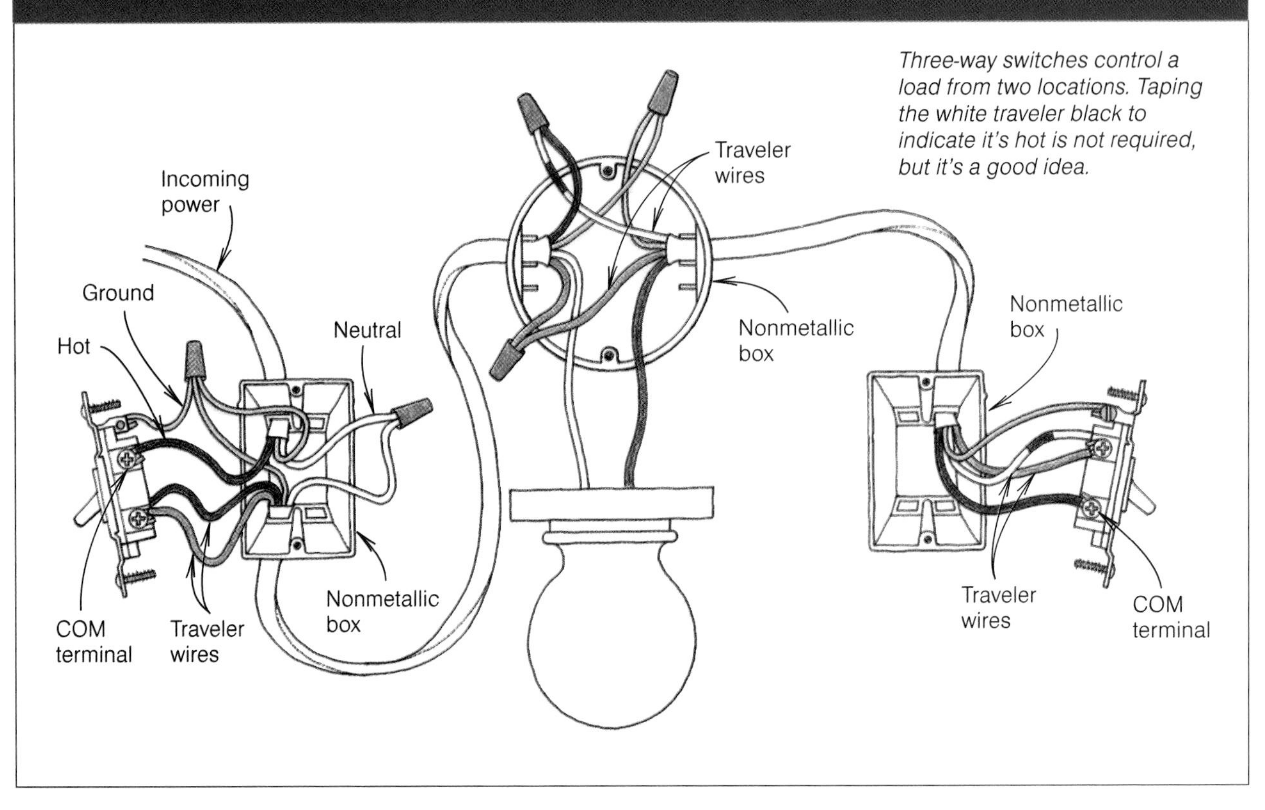

If you are replacing a three-way switch with another, be careful when transferring the connections. Not all switch manufacturers put the terminals in the same locations. Double-check which terminal is the COM and which are travelers. The screw for the COM terminal is normally marked by black paint.

I've made many service calls where people replaced a single switch with a three-way switch but didn't transfer the wires properly. The most common mistake is getting all the wires confused. I've had several people wire direct shorts into the line. Throw the switch to turn on the light, and POW, the breaker goes off. Other times I've seen folks leave off the neutral. To be sure you make the transfer correctly, mark the wires with tape as you take them off. First mark the COM wire and then mark the two travelers.

Once wired, the switch must be checked. You have to throw the switch three times to check it. The first should turn on the lights at one location. The second test is to turn them off at the other end. The third test is to turn them back on again at the first location.

Dimmer switch

The dimmer switch was developed to provide continuous control over lighting as opposed to a simple on-and-off switch. A dimmer can do this by limiting the amount of time that current flows to the load. Three distinct advantages of a dimmer switch are that it saves electricity, extends the life of a bulb's filament, and provides the proper mood for a room. (Fluorescent lights require special ballasts and dimmers.)

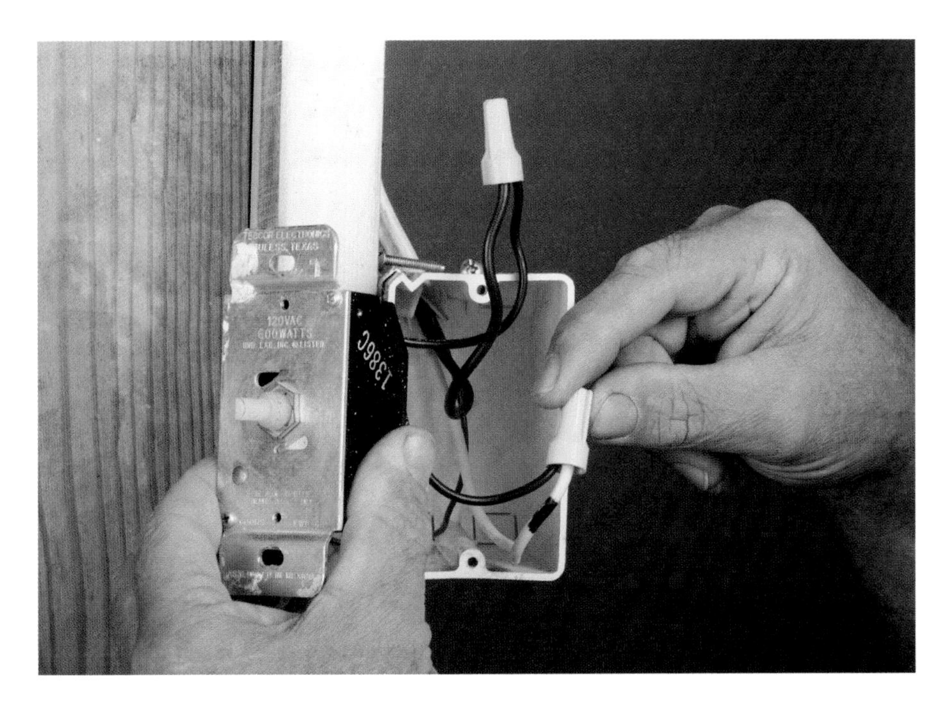

To wire a dimmer, splice the incoming black and white wires to the dimmer's wires and attach the ground wire to the grounding terminal, if any.

A standard dimmer switch is rated only for 600 watts (six 100-watt light bulbs) as indicated by the figure immediately above the knob. Never exceed this amount.

A dimmer switch is wired the same way as a single-pole switch. First, make sure the power is off. The dimmer has two black wires already attached to it. Simply splice the incoming black and white wires to these two wires (polarity does not matter). Then, if it needs it, attach the ground to the switch (not all dimmers have ground wires). That's it. If you're replacing a single-pole switch with a dimmer, simply remove the two wires from the single-pole switch and splice them to the dimmer's two wires (see the photo above). Then it's just a matter of pushing everything back into the box, turning on the power, and testing the switch.

Installing a three-way dimmer switch on a three-way lighting circuit is a little harder, but it's basically the same as a standard three-way circuit. However, only one of the two switches in a three-way circuit can be dimmed with a standard dimmer. The other must be left a switch. (Special dimmers are made for dimming both locations.)

Here's how to do it. Make sure the power is removed from the circuit. A three-way dimmer switch has the same number of terminals as a standard three-way switch—one COM terminal and two travelers—but it has wires already attached. Find the COM terminal wire and splice the incoming hot wire (black or red) to it. (If you are replacing a three-way switch with a three-way dimmer, simply remove the wires from the standard switch and transfer them to the dimmer wires.) Then the travelers can be spliced to the dimmer travelers. Again, there is no traveler polarity involved, so it doesn't matter which wires go to which traveler terminals. Once wired, push the switch into the box, turn on the power, and test the switch.

A couple words of caution. Standard dimmers are rated for up to 600 watts. Do not install a standard dimmer on a light circuit that will exceed that amount (see the bottom photo on p. 47). Also, it's normal for a dimmer switch to run hot, so don't worry if the front plate feels warm—it's also the dimmer's heat sink, designed to dissipate the heat generated by the dimmer.

The lights on this plug-in tester indicate that the receptacle is wired properly.

The heat is coming from the electronic components in the dimmer as electricity flows through them. Unfortunately, these components can be destroyed by this heat, so the heat sink (heat dissipater) is incorporated within the design. This is where the limitation of the 600 watts comes from. The incorporated heat sink physically can only dissipate up to that amount of heat. If you want to use higher-wattage bulbs, get a dimmer with a bigger heat sink (available at most electrical distributors). Beware—some designs hum.

TROUBLESHOOTING

If you are having problems with your newly installed receptacle or switch, there are a few ways to check whether it has been wired correctly or if there is some other problem. Some troubleshooting requires only a little common sense to solve obvious problems. Other situations will require some special tools, such as a multimeter or a plug-in tester, to solve the not-so-obvious problems (for more on inspecting receptacles and switches, see Chapter 2).

Receptacles

Receptacles are easy to troubleshoot. First look at the obvious. If you're having problems, check to see if the power is on. Then turn off the power and physically check to see if the receptacle broke apart during installation. Next, make sure all connections are tight (if you used the push-in terminals of a cheap receptacle, loose connections are a common problem).

If these are not the sources of trouble, use a plug-in tester to check if the receptacle has been miswired (see the photo on the facing page). The two most common problems are switching the hot and neutral wires and leaving off the ground. Both problems are indicated on the tester's lights and are easily fixed by reversing the wires or connecting the ground wire.

Switches

The first troubleshooting procedure for a suspected faulty switch is to check the obvious: Note how it works as it is switched. Many times a faulty switch just doesn't feel or sound right. Wiggle the switch while it's in the on position and see if the light flickers. If this is the case, the switch may be physically broken or worn out and will need to be replaced.

Though the problem could still be the switch, never overlook a bad light bulb or light fixture. I've made many service calls to repair lights—a few to my local courthouse—and the only problem was bad bulbs (an electrician charges a lot to change bulbs). Check the fixture with a bulb you know is good.

To check the bulb, you can shake it to hear the broken filament. But this is not always the best test. Use a continuity checker or screw the bulb into a known good lamp. If the bulb is good, check the center contact inside the faulty light's socket. It could be excessively bent over and not able to make contact with the bulb's center contact after years of use. Make sure the power is off, use a pencil eraser to clean the contact, and then, with a small screwdriver, bend it up slightly (⅛ in.). So much for the obvious.

The not-so obvious, however, is what troubleshooting is all about. To determine if a common single-pole switch is working properly, simply measure the voltage across its two terminals. This can be done with the switch still in the box. With the cover plate removed, place one probe on one screw and the second probe on the other, being sure not to short out the probes on the sides of a metal box. The voltage should be around 120 volts when the switch is off and close to zero when the switch is on.

If the power has been removed from the entire circuit, use the multimeter as a continuity tester to check the switch. A standard single-pole switch should have continuity (a closed circuit) across its two terminals when the switch is on and no continuity (an open circuit) when the switch is off. Always take the switch out of the box and remove one of the two wires on the switch's screw terminals before testing to eliminate any false readings.

If the switch is powering a light, you can measure the voltage from the light's center contact to the screw thread base of the light. If it measures 120 volts when the switch is on and 0 volts when the switch is off, the switch is good.

To check a three-way switch, first make sure the power is off, then check the continuity from the COM terminal to each of the two other traveler terminals. One side should always read continuity, and the other should not. It doesn't make any difference which side reads what. If the switch you're testing is the one that has power applied to the COM terminal, it can be checked by measuring power first on one of the traveler terminals (from ground or neutral) and then on the other as the switch is thrown.

4

INTERIOR LIGHT FIXTURES

The most common mistake people make when choosing a light fixture is picking the one that simply looks best. Just because a fixture is attractive in its setting, it doesn't mean you have made the right selection.

Although looks are important—the fixture should match the style of the room, fulfill its purpose, and be attractive—there's more to choosing a light fixture. You must also consider all the problems a particular fixture can have.

The purpose of this chapter is to provide you with enough information about light fixtures that you buy not only the one you find attractive but also the one that best serves your lighting needs without future problems. I'll talk about how to choose, install, and troubleshoot incandescent lights, recessed lights, track lights, and fluorescent lights. Because some of these fixtures require installation of an electrical box in the ceiling, let's begin the discussion there.

CEILING BOXES

Picking the right electrical box for a light fixture (called a ceiling box) is important. If the wrong box is used, the light fixture could fall or the wires could become damaged from overcrowding. Ceiling boxes are available in metal and nonmetallic designs, and even though the metal boxes can hold more weight, I prefer nonmetallic simply because of the safety factor of having a nonconductive box. I've seen too many wires being cut by the sharp edges of a metal box and shorting out (metal boxes are more of a fire hazard, too).

Nonmetallic rectangular receptacle and switch boxes are different than ceiling boxes (for more on receptacle and switch boxes, see Chapter 3). The screws that hold the fixture onto the ceiling box are larger than those used for a receptacle or switch box and provide better holding power. If you hang a light fixture from a receptacle box, the light could fall onto the floor or someone standing under it.

Ceiling boxes differ by the way they mount and by their volume (cubic inches of space inside the box). As with a receptacle or switch box, always use the largest-volume box available. But not all light fixtures utilize a ceiling box. Some fixtures, such as fluorescent and recessed lights, will not need boxes because they already have a splice box in their framework or the fixture was supplied with a whip (long wires surrounded by flexible steel) that goes to a splice box several feet away from the fixture.

Ceiling Boxes

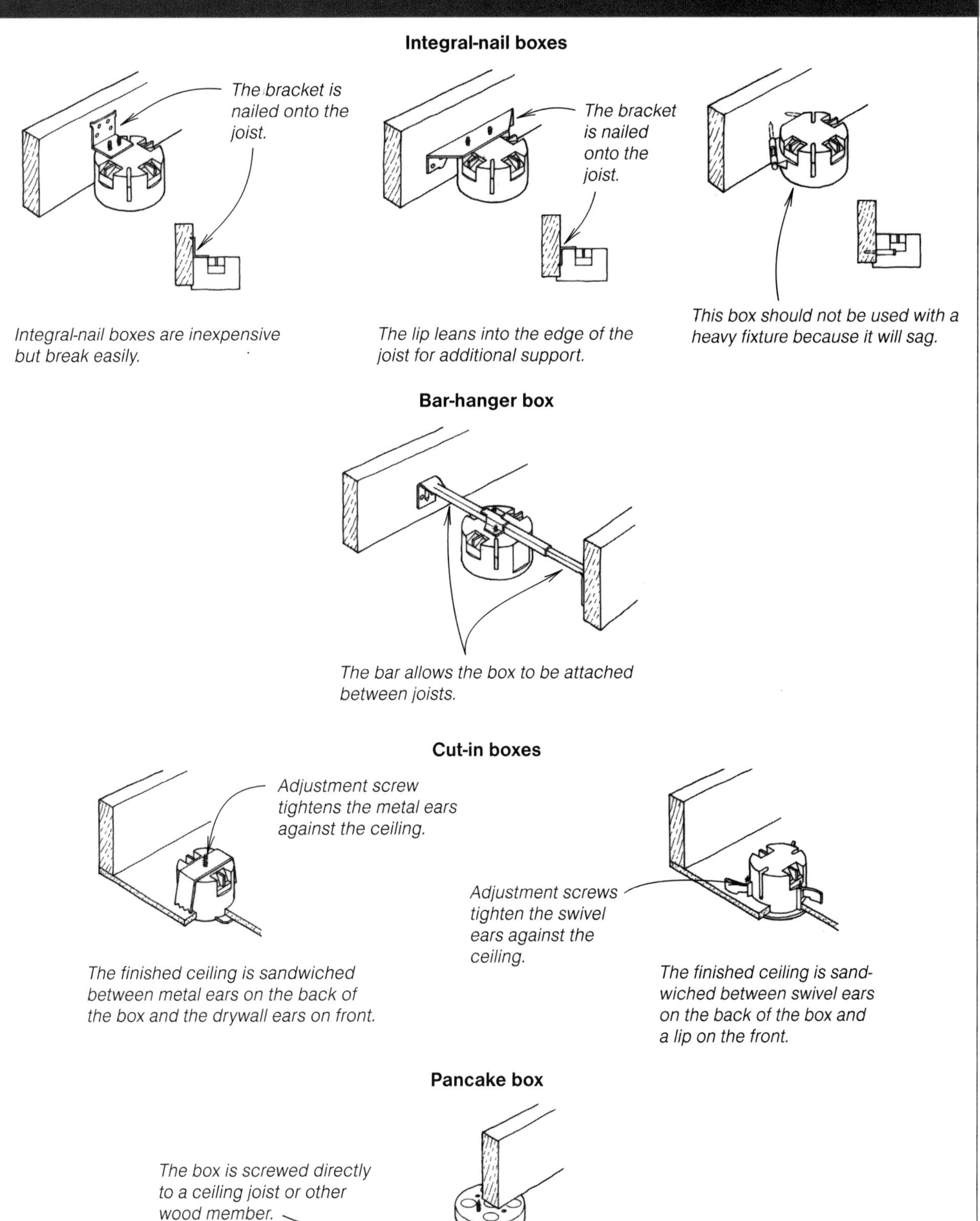

There are several types of ceiling boxes, the most common being the integral-nail box, bar-hanger box, cut-in box, and pancake box (see the drawings on p. 51). Most ceiling boxes, regardless of design, have volumes ranging from around 13 cu. in. to 23 cu. in., but pancake boxes have less (usually 4 cu. in. to 6 cu. in.).

The simplest and least expensive box is the integral-nail box. This type is easy to find, but it's cheap and breaks easily during installation. A better-quality integral-nail box is one with brackets to hold the box in place, which provides better support for the light fixture.

A bar-hanger box can be used in both new construction and renovation work. This type of box has a 16-in. or 24-in. bar that attaches between ceiling joists, and you can slide the box to any location along the bar. This type of box tends to sag under the weight of a fixture with even moderate weight, though.

A cut-in box is used for renovation work and makes it easy to install a light fixture. It is attached to the ceiling by sandwiching the finish ceiling between the drywall ears on the front of the box and a locking mechanism on the back. A screw tightens the locking mechanism against the back of the ceiling, locking the box in place. Unless a cut-in box is sandwiched to solid wood, it should only be used with a fixture that is very light in weight.

A pancake box is versatile because it can be used in both new construction and renovation work, and it provides sturdy support for a light fixture. For best results, attach the box directly to a stud or joist. The biggest drawbacks to a pancake box are that it's metal, so it needs to be grounded, and it is thin (usually about ½ in. thick) and cannot accommodate a large volume of wires. Its biggest benefit is that it fits nicely under the light fixture's base. Caution: 3½-in.-square boxes (4 cu. in.) are sold that cannot comply with code. The minimum-size box you should use is 4 in. square (6 cu. in.).

Mounting a ceiling box is simple. It is either screwed or nailed directly onto one joist or between two of them. A light fixture that weighs 50 lb. or less can be supported by the ceiling box alone. A fixture that weighs more than 50 lb. is required by the NEC to be secured into the house framing. However, I suggest not getting close to that weight.

INCANDESCENT LIGHTS

The most common light fixture in the home is the incandescent fixture. There are a wide variety of styles, and which you choose will depend on its purpose and what you find attractive. But it's important to choose the best-built, best-designed fixture to ensure that it will last a long time without trouble.

Choosing the right fixture

One of the first things to look for in an incandescent light is an air gap between the light's cover and the base of the fixture. The heat generated by the light bulb has to go somewhere, and without this air gap to allow heat to escape, it will move up into the fixture base and then into the wiring (see the left photo on the facing page). I've lost track of the number of light fixtures I've seen with burned and cracked wiring, along with blackened and deformed wire nuts at splices.

Beware of light fixtures with those beautiful teardrop bulbs, typically found in kitchen or dining-room lights (see the right photo on the facing page). Oftentimes these fixtures have a plastic or cardboard sleeve that covers the base, but it blackens and cracks over time from the heat generated by the bulb. You

Poorly Designed Light Fixtures

The light fixture above has no air gap between the cover and the base of the fixture, so heat generated by the bulb is trapped. Look for a fixture that allows the heat to escape.

Teardrop bulbs are pretty but will only stay so for a short period if the base around it is poorly designed. The heat from the bulb has blackened and cracked the plastic base on this fixture.

either have to cut the sleeve back or buy a cover for it (which could also crack and blacken).

Also make sure the fixture has standard screw-in bulb sockets, not the smaller, fancier specialized sockets. Bulbs with the standard base are less expensive and are stronger that those with the smaller base (see the top photo on p. 54). I've had many bulbs with small bases break at the metal-to-glass connection as I screwed them into the fixture. The fixtures with special bulbs are typically trendy designs, and as such, change or disappear altogether over time. That means you may not be able to get identical bulbs for them. So if you choose a fixture with unusual bulbs, be sure to have plenty of spares on hand because they may be hard to find in a few years. This happened to me. Above my bathroom vanity I installed strip lights with special iridescent bulbs. Now, because I cannot find this type of bulb, the fixture does not have the same effect as it used to.

Note the different base sizes in these bulbs. The standard size on the left is stronger and easier to find. The smaller base on the right has a large-diameter bulb and sometimes breaks at the bulb-base connection as the bulb is screwed in.

Make sure you do not insert a bulb that exceeds the wattage rating of the fixture. The rating is usually stamped inside.

If you are installing a fixture in a wet area, such as a shower, make sure that it can be used in that area (look for the UL label saying it is okay for use in wet areas) and that it is powered by a GFCI.

Another obvious but often overlooked quality for a light fixture is the ease with which you can change the bulb or clean the fixture's cover. Remember that the fixture will, for the most part, be mounted on a ceiling. A fixture that does not make it easy to change the bulb can be frustrating. I know one design that makes it very difficult to gain access to the bulb. You have to open a little door to get your fingers in, then you bend back small metal tabs to remove the glass sides. This is not an easy job while standing on a stepladder.

One last consideration—but not last in importance—is the wattage of the fixture. Most incandescent lights will have a maximum bulb wattage of 60 watts for one to three bulbs. This is enough illumination for general lighting but not enough for reading. If you need more light, either mount more than one fixture or search for one listed for higher wattages. Do not install a bulb in a fixture that has a higher wattage than the maximum wattage of the fixture (see the bottom photo at left). For example, don't put a 100-watt bulb in a fixture with a maximum wattage of 60. This could generate enough heat to melt the insulation on the wires and become a fire hazard.

Installation and wiring

When installing any light fixture, be sure to follow all manufacturer instructions and pay attention to all warning labels. Ignoring instructions or warning labels could be dangerous and will result in a light that does not work properly. Also be sure you have the correct size box for the fixture.

The first step is to remove power from the circuit. If you are installing a new light fixture, as opposed to replacing one, run the cable to the fixture's location (for more on running cables, see Chapter 3). If you are installing a new circuit from the main panel to the light fixture and have never run cable through walls, you might want to call an electrician to do the job. You can then make all the necessary connections.

The installation of a light fixture can go smoothly by thinking ahead. First, make sure you have the fixture in hand before you begin. I can remember one mistake I made when I installed a wall-mount light fixture adjacent to an exterior door. I mounted the box and pulled all the wiring through, then picked up the fixture. When I tried to install it, however, I realized it was too big—the top of the light hit the soffit. I try not to make the same mistake twice. Now I always make sure I have the fixture in hand so I can see how it fits before I begin any work.

Once you are sure the fixture fits, and after double-checking to see that power is removed from the circuit, begin mounting the fixture. A small flat bar (called a mounting strap) attaches to the ceiling box, and the fixture attaches to the bar (see the drawing at right). If the fixture is heavy, some other support may be required. The instructions should tell you so.

Wiring is pretty simple. The hot incoming black wire connects to the fixture's black wire, the white neutral connects to the fixture's white wire, and the ground wire connects to the flat bar or green wire of the fixture. Be sure to ground the ceiling box too if it is metal. That's it. After wiring is completed, restore power, insert a bulb, and see if the light works.

Wiring an Incandescent Light Fixture

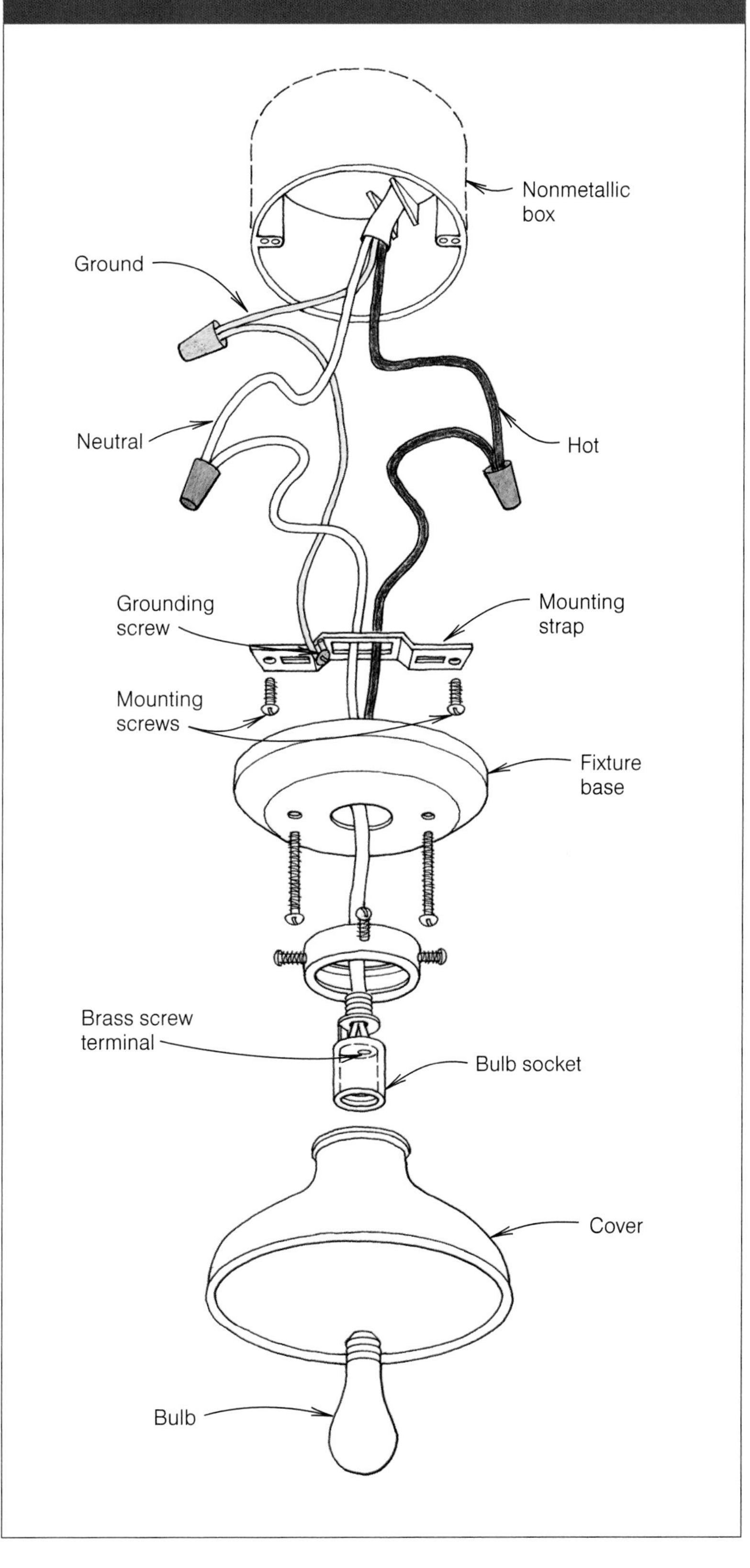

RECESSED LIGHTS

Recessed lights (also called can lights) can give a variety of illuminating effects throughout the house. Because they are inset into the ceiling, they are unobtrusive, which also makes them perfect for rooms with low ceilings. Recessed lights are commonly used for task lighting in kitchens and closets and for accent lighting in other areas of the home (say, to highlight works of art). There is no need for a ceiling box in this installation. The fixture is attached directly to the ceiling joists.

Choosing recessed lights

The main factor in your choice of recessed lights will be the size and type of housing. There are two basic types: IC type and non-IC type. A recessed light with an IC-type housing allows insulation to be around and over its mounting base in the ceiling or attic. A recessed light with a non-IC type housing must be kept at least 3 in. from any insulation. If insulation is touching or too close to a non-IC fixture, the heat generated by the light could start a fire (remember this when storing cardboard boxes in the attic). Housings are also available for special applications, such as sloped ceilings, closet lights, shallow spaces, and high-wattage bulbs (up to 200 watts).

In general, IC-type recessed lights have lower maximum wattages than non-IC types (see the photos on the facing page). The wattage ratings are labeled inside the fixture. As with any light fixture, do not exceed the wattage of the light. Both types of recessed lights have a heat sensor that will cause the light to cut off if it gets too hot or if its maximum bulb wattage is exceeded.

When choosing a recessed light, be aware of how easy it is to replace the bulb—in some models, it's a hassle. Be sure the design allows you to get your fingers in to remove the bulb. If it does not, you may have to buy a special suction-cup type of tool to remove the bulb. I personally would much rather remove it by hand. It's simpler that way.

One other note, before running the wiring, make sure the fixture you have chosen will fit. Some ceiling cavities may be so shallow that you must buy a shallow recessed fixture.

Installation and wiring

Once you have chosen the housing and fixture you like—and one that will fit the space in your ceiling—you can begin the installation process. As usual, make sure power is off and run the wiring first. Also be sure to follow all manufacturer instructions and warnings. There are basically two steps in the installation process: mounting and wiring.

Most recessed lights are mounted on two sliding arms that are nailed to the ceiling joists. Be sure to remove any insulation in the way and position the fixture so that the finished wall will be able to slide under the frame and butt up against the circular housing at the bottom.

Wiring to the light is done inside a metal splice box mounted on the light's frame. Inside the splice box will be a black wire, a white wire, and a ground wire. The incoming power cable enters the splice box via one of the many knockout holes. Using a screwdriver, pry away the knockout center and install an NM connector. An NM connector (see the photo on p. 99) is installed in a knockout in the box to keep the cable away from the metal box's sharp edges and to tighten down on the cable to keep it from being pulled out of the box.

Insert the cable through the connector and splice black to black, white to white, and ground to ground (see the drawing on p. 58). Gently tighten down the two

Two Types of Recessed Housings

All recessed light fixtures have their maximum wattage labeled inside the housing. They also have heat sensors that cause the light to blink or cut off when the temperature gets too high.

This IC-type recessed light has a maximum wattage of 75, and the heat sensor is mounted on the bottom of the housing.

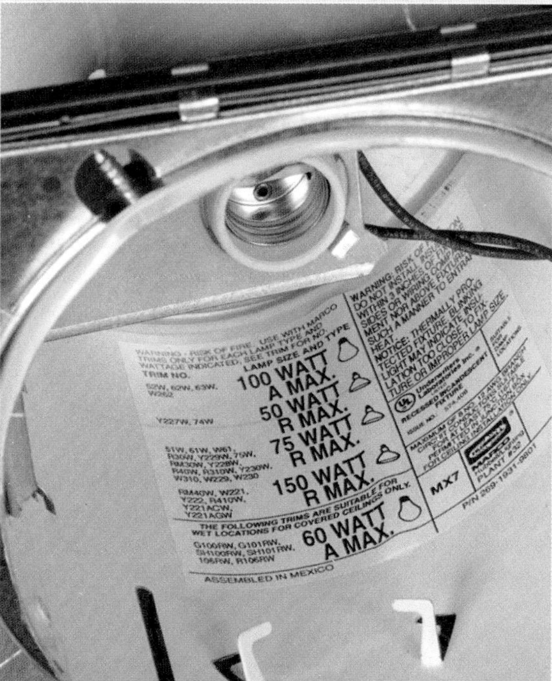

This non-IC-type recessed light has a maximum wattage of 150. The heat sensor is mounted on the side of the housing out of view.

Wiring a Recessed Light

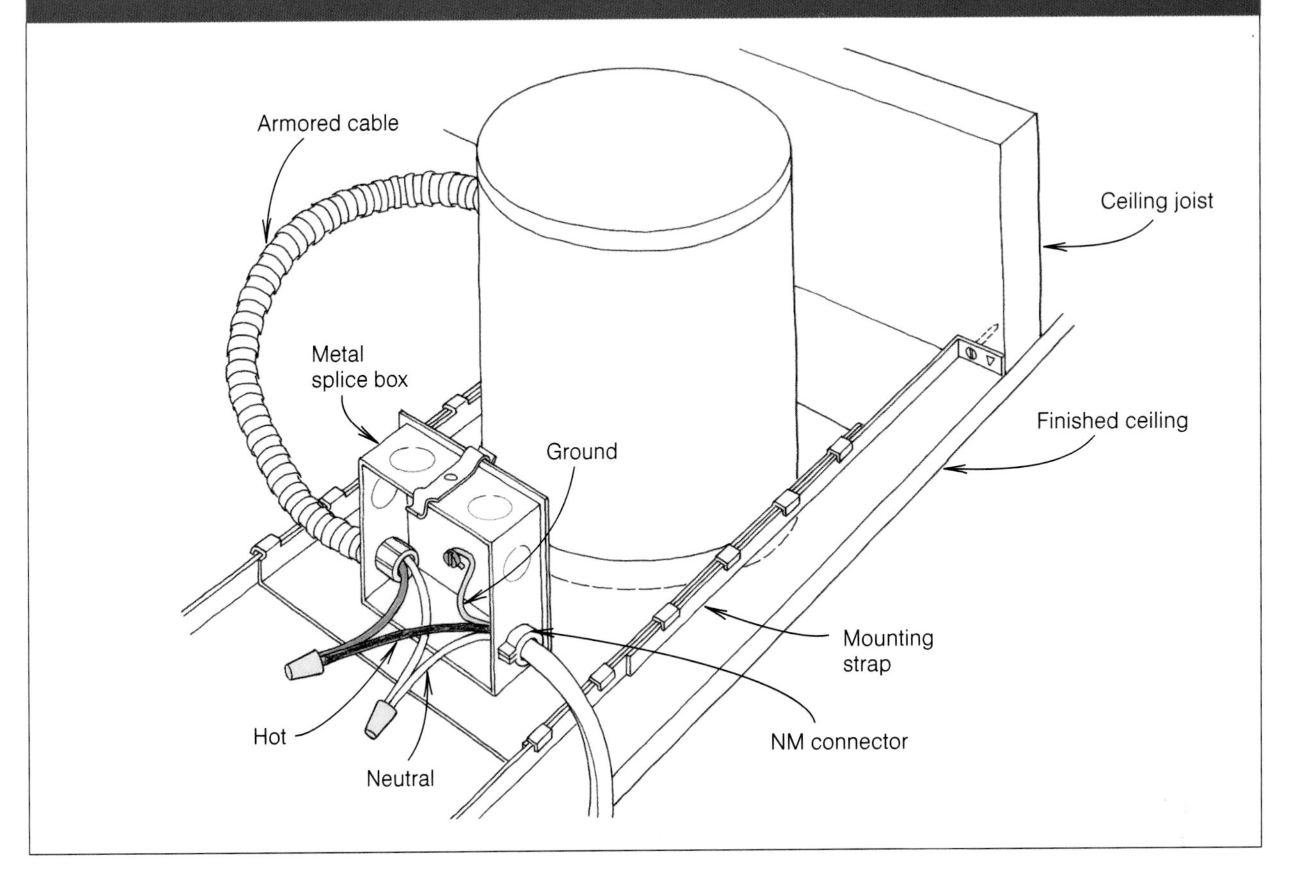

screws on the NM connector to hold the incoming power cable in place. (I say gently because I once had a helper tighten the screw so much that he cut through the insulation on the wires and shorted them out.) Then restore power and test the light.

After the fixture has been wired and is working properly, you can replace the insulation you removed earlier. Remember, if you have a non-IC fixture, keep the insulation at least 3 in. from the housing (it helps to build a 1x frame around the fixture to hold the insulation away). It is preferred, and some model energy codes require, that IC fixtures be used wherever insulation will be, such as in attics.

TRACK LIGHTS

Track lights are typically used to provide accent lighting. What's great about these lights is that you can place a lamp at any spot on the track, but you only have to bring in one cable to power all the lights on the track (see the photo on the facing page). Another good thing about track lights is that they can be mounted on any flat surface, such as ceilings, walls, or even cabinets.

Track lights allow you to insert lamps into the track at any location. When locked in place, the two bare copper tabs on the bottom of the lamp (see inset) make electrical contact with the track.

Choosing track lights

In general, the type of track lights you choose will depend on your personal tastes and on your space requirements. But there are a few other things you have to consider as well, such as color, lampholder design, and what lighting effects you want.

Track lights are available in a wide variety of colors (although the selection seems to be dwindling), with black and white being the most common.

When picking a track-light design and manufacturer, pay particular attention to the housing design. I prefer track lights that have an air gap between the housing and the bulb to allow heat to dissipate. The air gap also allows you to get your fingers in to remove the bulb. It's also best to buy track lights that accept only standard flood-light bulbs or spot-light bulbs. These bulbs are easier to find and are less expensive than specialty types.

As you shop around, also remember to think about what kind of lighting effects you want to achieve. Doing so will help you pick the right housing and bulb for the job. There are three types of housings. The first is the standard tube

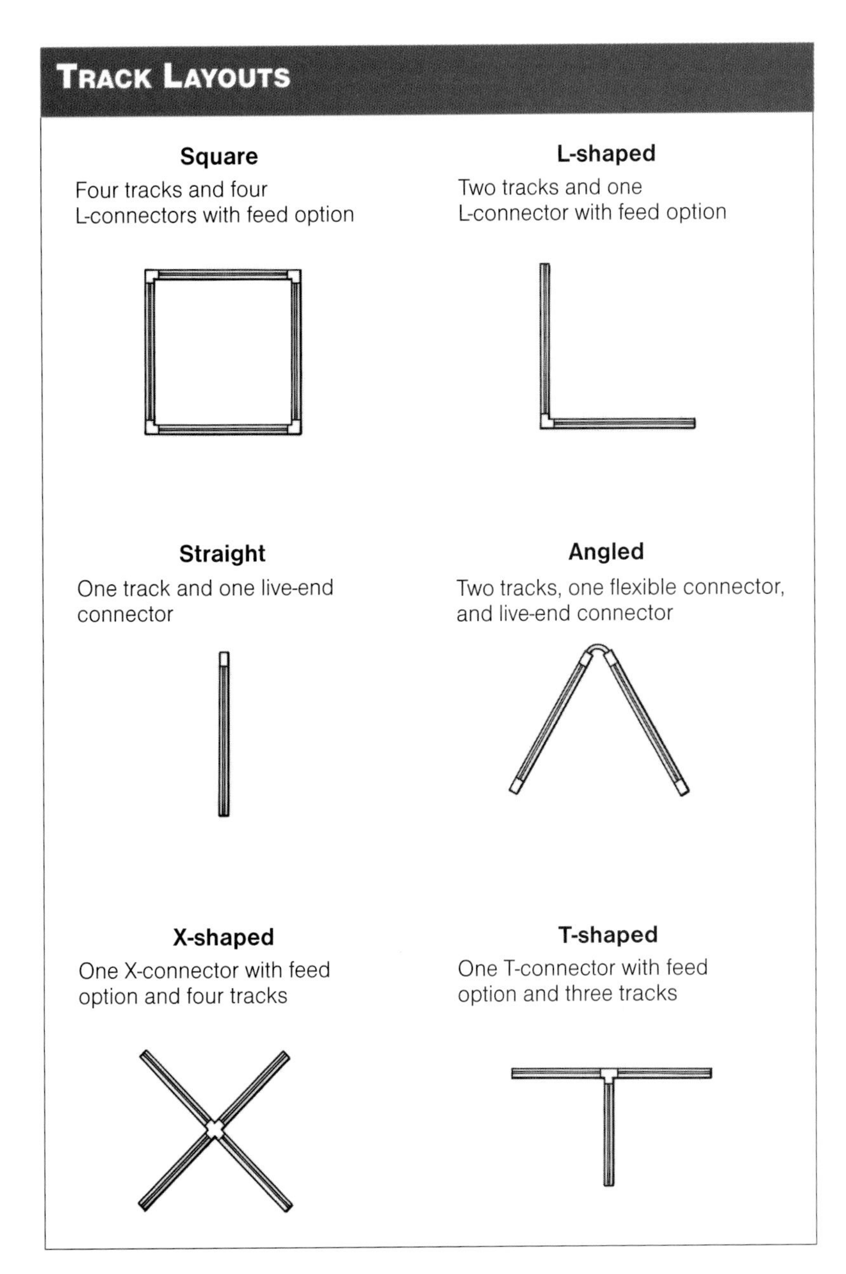

housing most common in homes and is used for general-purpose or task lighting. The second type of housing is a reflector housing, which is used to spread light over a large area—perfect for work areas. The third type of housing is the wall wash, which is used to accent a specific wall area—to highlight a work of art, for instance.

The biggest drawback to track lighting is that you can add lamps that could have varying wattages—and the watts add up quickly. To be safe and to avoid overheating the wiring, I always install track lights on a 20-amp circuit (12-gauge cable protected by a 20-amp fuse or circuit breaker).

Track layout

Track lighting is versatile because the tracks can be arranged in a variety of ways.

Track sections are normally made of aluminum and are available in 2-, 4-, and 8-ft. lengths. They are wired internally with 12-gauge conductors, and the sections can be arranged in any order imaginable (see the drawings at left): square, straight, angled, or in the shape of an L, T, or X. The tracks also come with a wide variety of connectors to accommodate every layout (see the drawings on the facing page). The connectors serve two purposes: to attach track sections and to bring power to them. When sections are fitted with flexible connectors, they can form almost any angle.

And if you want to have two different accent lighting effects in the same area, track sections are available with two independent circuits, each one being controlled by a different switch. This setup allows you to control the lighting effects according to your mood.

Standard tracks must sit on a flat surface. If a wall or ceiling is not flat, the tracks will reveal gaps underneath. But that does not mean you cannot install track lights. You can buy special track-light fixtures that hang from the ceiling.

Installation and wiring

As with other light fixtures, it's important to follow the manufacturer's instructions and warnings when installing track lights. Also remember to turn off power before beginning.

Track-lighting installation must be thought out well in advance because the feeder cable from the main panel must be brought to the *exact* location the track begins or ends. To bring the cable to the right location, first lay out the track(s), then mark where the feeder cable will connect to the track. (Power is typically brought into the track through its end or through one of the connectors.) At your mark, drill a hole large enough to accept the track connector, then run the feeder cable from the main panel (for more on running cable, see Chapter 3).

Once the cable has been run, attach the track to the ceiling or wall. If you are putting the track on a solid surface, such as wood, you can screw the track directly to it. But if the surface is drywall, you'll have to screw the track through the wall into studs or ceiling joists. If needed, molly anchors can be used to attach the track to the drywall.

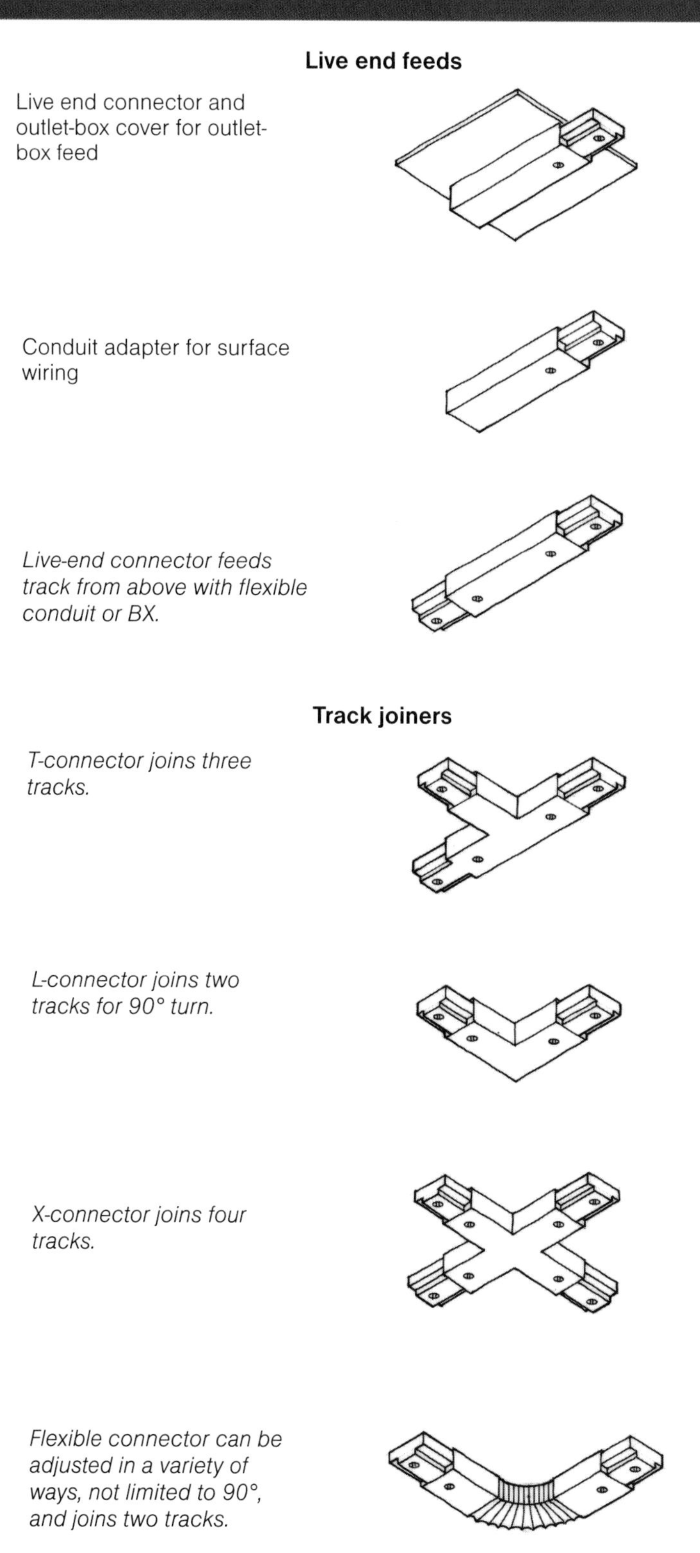

To wire a track, remove the cover from the connector and attach the white wire to the silver terminal, the black wire to the gold terminal, and the ground wire to the green terminal.

With that complete, make the connections. I find it's easier simply to connect the wires directly to the track terminals, but if there's more than one cable, you have to install a splice box separate from the track. The wiring is pretty basic. Remove the cover from the end or the connector where the cable is entering and attach the white wire to the silver screw terminal, the black wire to the gold screw terminal, and the ground wire to the green screw terminal (see the photo above).

Once the connections are made, test the lights. Insert a lamp into the track and lock it in place (most lamps lock by twisting them 90°). Restore power and turn on the switch. Once you are sure the system is working, install the rest of the lamps.

FLUORESCENT LIGHTS

Fluorescent lights are a very efficient method of illumination, which is one reason why they are popular. A fluorescent light provides two to four times the amount of light that an incandescent light provides, and the bulbs generally last a long time.

Choosing fluorescent lights

Your choice of fluorescent light will depend on what kind of lighting you need and where it will be located.

If the light is to be located in a workshop or garage, where appearance is not that important, buy a fixture without a cover over the bulbs. It's not very attractive, but it works and is less expensive than a fixture that comes with a cover. If you do opt for the cover, be careful with it when replacing bulbs because the plastic breaks easily.

The most common fixture lengths are 4 ft. and 2 ft. These accommodate bulbs with two pins on each end (see the photo on the facing page). For larger rooms, you might need 8-ft. fixtures. The bulbs for the 8-ft. fixtures have a single pin on each end. Be careful when installing these bulbs because they are unwieldy and break easily.

For extra light, high-output and very-high-output fixtures and bulbs are available. These fixtures are perfect for a work area or rooms with high ceilings. High-output fluorescents have different pin arrangements so that their bulbs cannot be accidentally installed in standard fluorescent fixtures.

Make note that some fluorescent fixtures come with plug-in cords, which are great for a quick and easy installation. Simply install a 120-volt receptacle near the light. For convenience, power the receptacle through a switch so that you can turn the light on and off with it.

The ballast

Problems with a fluorescent light, such as flickering or the light not working, may be caused by the ballast (which is a transformer inside the fixture). The ballast is a sensitive component that raises the voltage high enough to arc across a filament inside the bulb, which then excites the gas inside the bulb and produces light.

A standard ballast won't work well in cold areas, such as garages and outbuildings. If you want to put a fluorescent light in a cold area, make sure you replace the standard ballast with one designed for cold weather (this type of ballast is expensive).

A standard ballast also overheats easily, which will shut the light off. On one service call I had to drill holes in the fixture to allow heat to dissipate so that the light would stay on. If the ballast gets too hot, it could melt, leaving a black gummy residue inside the fixture.

A ballast is also the source of any annoying buzzing and flickering that sometimes occurs with a fluorescent light. To remove the buzzing or flickering, you can replace the ballast with an electronic one, which may cost as much as or even more than the entire fixture itself. Added benefits of an electronic ballast are that it operates cooler—it rarely overheats—and is around 30% more efficient than a standard ballast, with double the life expectancy. To replace a ballast, simply remove the old one and splice like wires together—black to black, white to white. Just make sure the power is off first.

There are three types of fluorescent bulbs commonly used around the house. At left is a high-output bulb, at center is a standard 4-ft. bulb with two pins, and at right is an 8-ft. bulb with one pin.

Installation and wiring

As with all light fixtures discussed in this chapter, follow the manufacturer's recommendations and warnings when installing a fluorescent light. There are two steps to the installation: mounting and then wiring.

A fluorescent light must be mounted securely to a supporting member. It can be mounted by directly connecting it to overhead beams with screws or by hanging it with chains or wire. Never attach the light to the grid frame of a dropped ceiling. The weight of the light will pull the ceiling down. Once I was working in an office where all the rectangular light frames were resting on the dropped ceiling. One ceiling section came loose, causing a light to fall down. But that was not the worst of it: All the other lights attached to the first also fell like dominoes.

The second step is wiring, which is pretty simple. First cut off power to the circuit. Bring the cable (either from the main panel or from a receptacle or hot switch)

Wiring a Fluorescent Fixture

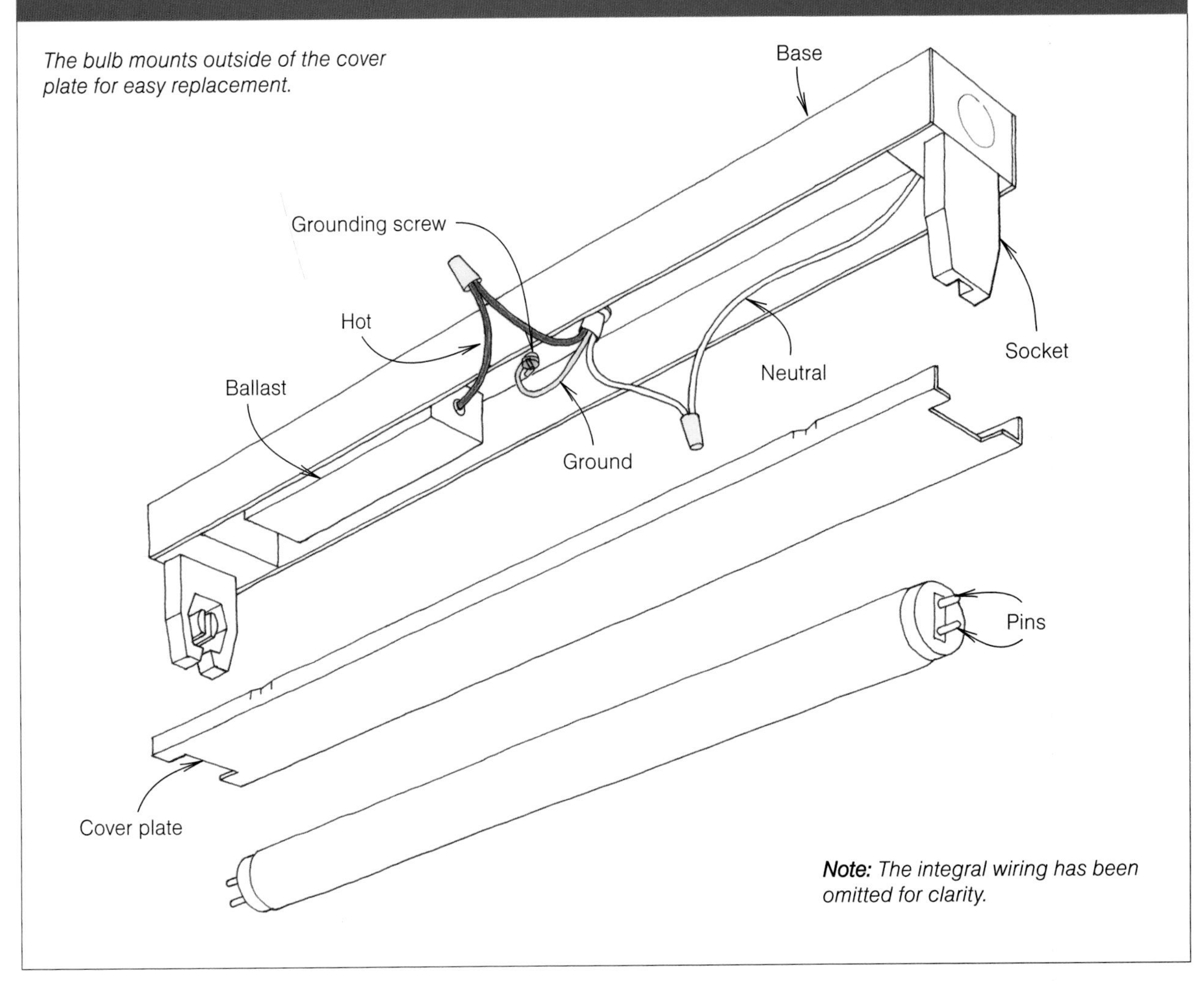

to the light. Then remove the aluminum plate that covers the internal wiring and the ballast.

A fluorescent light has internal black and white wires that connect to the internal wiring in the ballast and to the end connection. Simply bring the incoming cable through a knockout on the fixture and connect the incoming black wire to the internal black wire, connect the incoming white wire to the internal white wire, and attach the ground wire to the grounding screw (see the drawing above).

The biggest problem in wiring a fluorescent light is keeping the internal wires from falling out and getting in the way as you replace the cover plate. To alleviate this, I use electrical tape to hold the wires along the center of the fixture.

After replacing the cover plate, restore power to the circuit, insert the bulbs, and test the light.

Under-cabinet installation

A common installation for a fluorescent light is under a kitchen cabinet for close-up task lighting on the countertop or at the sink (see the photo at right). For this use, you'll need to buy a special low-profile light fixture that is around 1½ in. tall. The shallow height allows the light to be hidden behind the lip of overhead kitchen cabinets.

This type of light is screwed into the bottom of the cabinet, so don't use a screw that's too long. I did once and wound up screwing into a heavy wooden platter inside the cabinet. The wiring process is the same as previously described.

One last note. Under-cabinet lights cannot be wired into the countertop receptacles or even the kitchen, dining, or pantry receptacles. It's a code violation. Instead, this type of light should be tied into the kitchen lighting circuit. Some model energy codes require the fluorescent light to be switched separately from incandescent lights.

TROUBLESHOOTING

If your newly installed light is not working, or if you are having problems with an older light, most often there is trouble with the bulb or with the wiring of the switch or fixture, not with the fixture itself (for more on inspecting receptacles and switches, see Chapter 2), although older fixtures do have problems. To troubleshoot a light fixture, you'll need a multimeter. With any type of fixture, check the bulb first.

Incandescent lights

If you are having trouble with an incandescent light, there are a few troubleshooting procedures that can be done. First, replace the existing bulb with a known good bulb or perform a simple continuity test to check if the bulb is good. To do this, set the multimeter to check resistance or continuity. Touch one lead to the threaded base of the bulb and the other lead to the center contact on the bottom (see the drawing on p. 66). If there is little or no resistance (a reading of around 10 ohms) or the meter is buzzing, there is continuity (see the Glossary on p. 146), which means the bulb's filament is intact and the bulb is good.

A popular location for fluorescent light fixtures is under overhead kitchen cabinets. They are perfect for providing unobtrusive task lighting.

If the bulb is good, check the fixture next. If it is an older fixture, remove the bulb and check the center contact in the socket. Sometimes dirt or corrosion can build up on the contact. To clean it, turn off the power and rub the contact with a pencil eraser, being careful not to crush it down. Then turn on the power, replace the bulb, and check the light. Another problem with the contact is that it sometimes gets bent down from use so that it no longer touches with the bulb's center contact. To remedy this, turn the power off and use a small screwdriver to bend the contact up around ⅛ in. Don't bend it any more than that because the contact may snap off.

Testing an Incandescent Bulb

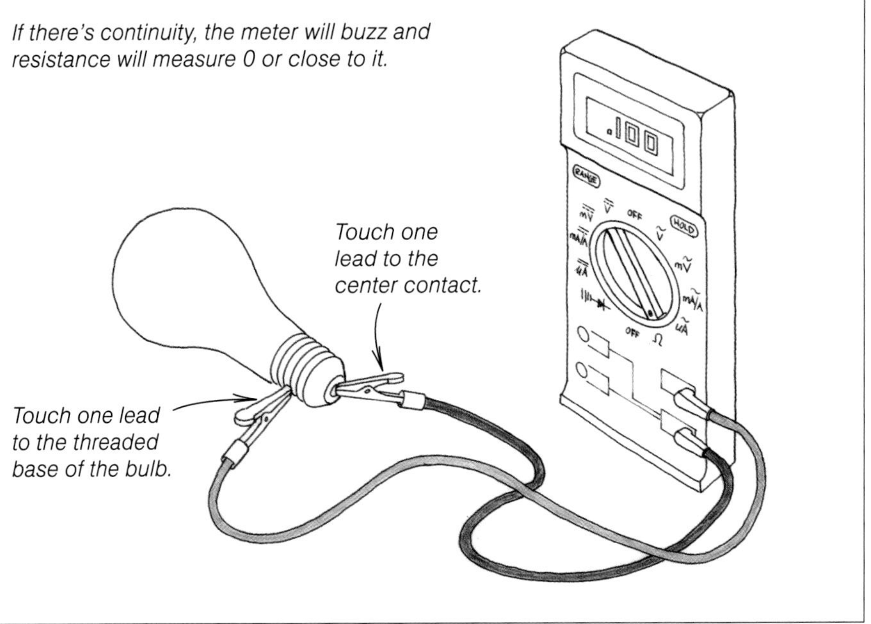

Replacing a Broken Bulb

Incandescent bulbs sometimes get broken while they are in the fixture, which makes them very difficult to remove. I've smashed a few in my time and have come up with some ways to get the bulb out without hurting myself.

First, make sure there is no power to the fixture by turning off the switch or breaker. (If you are working on an overhead fixture, wear safety glasses to keep glass from falling into your eyes.)

If any part of the glass is sticking down, the bulb can be removed with a potato. Slice a potato in half and push it onto the broken bulb and twist counterclockwise. If the bulb is not stuck, it should come out. If the bulb is stuck, grip the edge of the threaded base with a pair of pliers and turn to remove the bulb. If you can't get a grip on the base, bend it in with a screwdriver so that you can get the mouth of the pliers on it.

If the contact is not the source of trouble, check the switch wiring (see Chapters 2 and 3) and the fixture's wiring. If all of the wiring is good, the fixture is bad and will need to be replaced.

Recessed lights

As with incandescent lights, when a recessed light is not working, the first thing to check is the bulb. The procedure is the same as for an incandescent light. If the bulb is good, check the switch and fixture wiring to be sure there are no loose connections and to be sure the fixture is wired correctly.

If the bulb and wiring are good, the problem could be heat buildup. If there is too much heat, the light will blink or won't work at all. This is common when insulation has been placed around or too close to a non-IC type housing. If you find that this is the problem, remove the insulation immediately—it must be at least 3 in. from the housing.

Excessive heat can damage the insulation on the wires and even the socket. In some cases the wires and socket can be replaced—other times the entire unit will have to be replaced.

Track lights

There's not much to troubleshooting track lights because there is no internal wiring to check and overall they are pretty reliable. However, if you are having trouble with the lights, first check the bulb so see if it is good. If the bulb is not the problem, make sure the lamp is secured in the track correctly.

Fluorescent lights

Fluorescent fixtures have an assortment of problems—typically indicated by a flickering light—that are caused by either the bulb or the ballast.

First check the bulb. If a bad bulb isn't replaced and allowed to continuously flicker, it may ruin the ballast. First rotate or jiggle the bulbs to verify that the pins are properly inserted into the sockets. If that doesn't do the trick, check to see if the end sockets are broken or cracked. If they are intact, make sure they haven't pulled so far apart that the bulb's pins aren't making contact with the sockets. If all appears okay, try replacing the bulb with a known good one.

If the unit still doesn't work after replacing the bulb, remove the cover plate to expose the internal wiring. Use a multimeter to measure the voltage to the unit across the black and white wires bringing in power. If the meter reads 120 volts, the wiring is good, which means the problem most likely lies with the ballast.

Sometimes it will be obvious that the ballast has gone bad. Look for a black liquid seeping out of it. If you have to replace the ballast, make sure it matches the fixture. Replacing the ballast will cost about the same as a whole new fixture. Because the cost is the same, how to fix the problem is a question of labor. Do you want to spend a lot of time replacing the whole fixture, or would you rather spend a shorter period of time replacing just the ballast?

If the fixture is giving off a loud buzzing noise, you can try changing the bulbs. But you'll probably have to change the ballast, and even that won't guarantee against buzzing. The only ballast that will guarantee no buzz is an electronic one, which is expensive. (For more on ballasts, see p. 63.)

Some fluorescent fixtures have a starter, which is a small cylindrical can located in front of the unit. If the bulb is flickering or not working, it could mean that the starter has gone bad. To remove the starter, push it in slightly, then turn it counterclockwise.

If you are still experiencing problems after all these checks, verify that the wiring has been done correctly.

5

OUTDOOR LIGHT FIXTURES

You can buy heavy-duty, custom-made entry lights that will stand up to any abuse the weather throws at them. This heavy-duty fixture was made by Crenshaw Lighting of Floyd, Virginia.

My friend couldn't believe his eyes. He was proudly showing off the new motion detector he had just installed, but as I approached it, the light didn't turn on. I even walked right up to it, and the light still didn't go on. Although my friend wired the motion detector properly, he made the common error of mounting it in the wrong location for it to sense movement. It just goes to show that outdoor lighting requires more thought than just which wire goes where.

Wiring an outdoor light is pretty simple: bring the cable to the light's location, install a box, if necessary, and splice the wires (for more on running cable and splicing, see Chapter 3). What makes installing an outdoor light fixture tricky, though, is choosing the right light for the job—whether it's for security, for general illumination, or simply for aesthetics—and locating the fixture so that it does its job well.

Along with those considerations, you also have to think about the quality, initial cost, operating costs, the color of the light, and its maintenance (keeping it accessible for bulb changing). In this chapter I will discuss all of these issues, with the goal of helping you choose the right outdoor fixtures for your needs—whether they are entry lights, floodlights, motion detectors, low-voltage landscape lights, or post lights. I will also show you how and where to install each type of light for optimum performance.

QUALITY FIRST

According to Woody Crenshaw, a lighting expert and owner of Crenshaw Lighting in Floyd, Virginia, there are four things to look for when buying any type of high-quality, heavy-duty outdoor fixture. First, all metal parts should be made of nonferrous material (nonferrous means no iron) to keep them from rusting or corroding. Second, the bulb socket should have a brass or copper shell. Aluminum sockets tend to corrode, making the bulb almost impossible to get out. Third, the see-through material should be glass or UV-stabilized plastic—not Lexan or styrene. And fourth, solid brass is always better than plated brass.

Also, you can't just install any light outside. It must be watertight and approved for such. The two UL classifications for outdoor lights are wet and damp. Wet is for lights that will be installed in full weather—rain, snow, and sleet. Damp is for those lights that are outside but installed under some type of shelter like the eave of a roof.

Mounting Exterior Fixtures

It's standard procedure to install a wall-mounted exterior fixture on an electrical box—either within the wall or on the surface.

If you choose to install the box within the wall, the front of it must be flush to the siding surface. This is not difficult if you have flat siding, such as board and batten or T-111. But if you have beveled siding, it's not so easy—the angled surfaces of the siding won't match the flat surfaces of the box.

I prefer to mount exterior fixtures on boxes mounted on the siding itself because it's easier—you won't have to cut the siding around the box—but the same problem remains: how to deal with angled surfaces.

For some fixtures, such as floodlights located high on the house, it's okay if they sit at an angle because they aren't that visible. But for other fixtures, such as entry lights that are easily seen, mounting on an angle will look terrible.

In this situation, it may be advisable to hire a carpenter to make a fancy base for you that will allow the fixture to be plumb to the wall and flush to the siding.

Another option is to grab a piece of the beveled siding and invert it on the installed siding to make a flat area (see the drawing at right). If the house has vinyl siding, you can buy an adapter that will allow the fixture to sit flat on the exterior.

If you want to install a fixture on the surface of a masonry wall, you can easily drill a hole for the cable and mount a box using masonry anchors. But if you intend to install a flush-mounted box within the wall, you'll have to chip out the masonry so that box and wires will fit within the wall.

For some installations, such as outside incandescent floodlights, it may be appropriate to mount the fixture under the eaves. Simply bring the cable out of the soffit, install the appropriate box, and attach the fixture.

Making a Surface Level

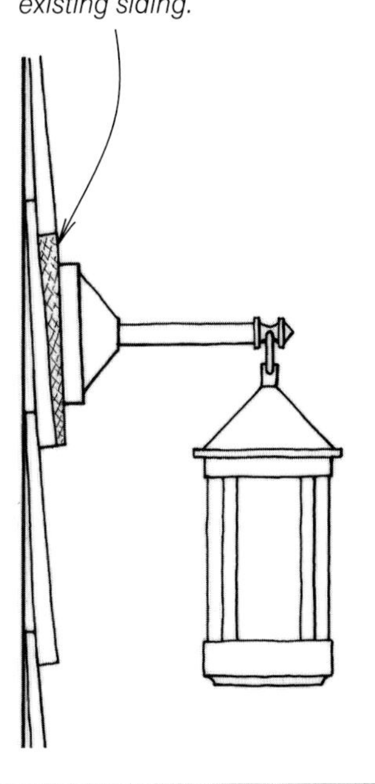

ENTRY LIGHTS

Entry lights are typically used to provide general lighting at a doorway. The type of fixture you choose will depend on your tastes and on the architecture of your house, *but it's important that you not choose the cheapest available*.

Because an entry light (or any other outdoor fixture) is exposed to the weather, you should check if there's a warranty on the finished surface. Some low-cost units will rust and tarnish after only a few years of exposure to the weather. I installed cheap entry lights near the sliding doors on my deck. After only two years, they rusted and looked like they went through a war. I replaced them with heavy-duty lights handcrafted from a custom lighting shop (see the photo on the facing page).

You don't have to spend a ton of money on custom lights, but you should spend the extra buck to get a quality fixture.

Installation

The trickiest part of installing an entry light is mounting it. You'll need a level surface, so if the building has beveled or irregular siding, you'll need to do some extra work (for more on mounting exterior fixtures, see the sidebar on p. 69).

Once you have a flat surface on which to mount the fixture, run the cable, pulling it through the exterior wall, and install the appropriate box (for more on boxes, see Chapter 4). Then, after verifying that power is off, simply attach like wires (black to black, white to white), ground the fixture, and mount the fixture. It's that easy.

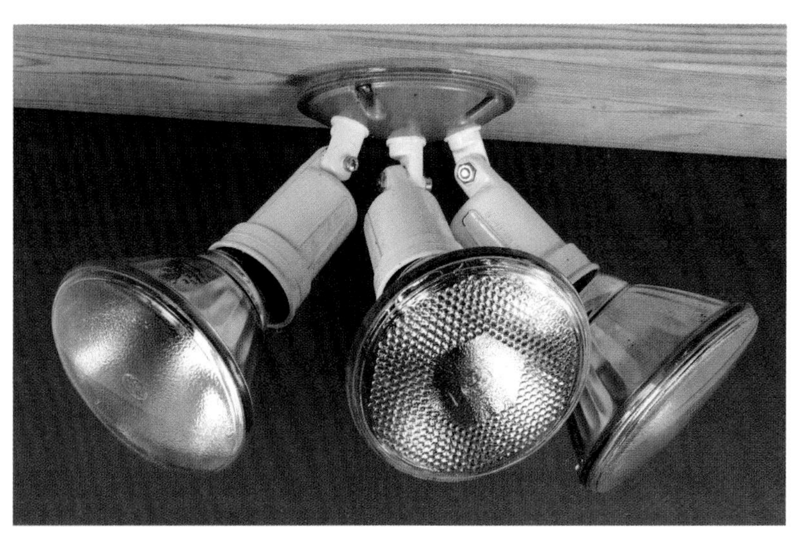

An incandescent floodlight can project light in three different directions for wide coverage or at one spot for bright, concentrated illumination.

Once the fixture has been installed, restore power and test the light. (If you are installing a switch for the fixture, see Chapter 3 for information on wiring it.)

FLOODLIGHTS

Floodlights are used to illuminate large, wide areas of the property. The most common floodlights on the market today are incandescent, quartz halogen, and high-intensity discharge (HID). The biggest differences between the three are the amount of illumination provided and the cost of the fixture.

Incandescent

Incandescent floodlights are the most commonly installed and can handle a variety of wattage bulbs, from 75 watts to 150 watts (see the photo at left).

Incandescent fixtures are normally equipped with one to three articulating arms that hold the bulb sockets (make sure you choose a fixture that accepts standard screw-in floodlight bulbs). The arms allow you to adjust each bulb to any angle so as to illuminate a large area or a smaller area. For example, if you were to mount a unit with three arms on a house corner, one bulb can light one side of the house, the center bulb can cover the corner, and the third bulb can cover the adjacent corner of the house. If you need a lot of illumination at one spot in the yard, all three bulbs can be positioned to shine on that location.

Incandescent floodlights are inexpensive to purchase, but they are more costly to operate than quartz-halogen or HID floodlights because they draw more current. However, if they are on only for short periods of time, such as to illuminate a parking area for company, the cost of operating them isn't significant.

Incandescent floodlights have other drawbacks besides their operating costs. Most of the problems have to do with

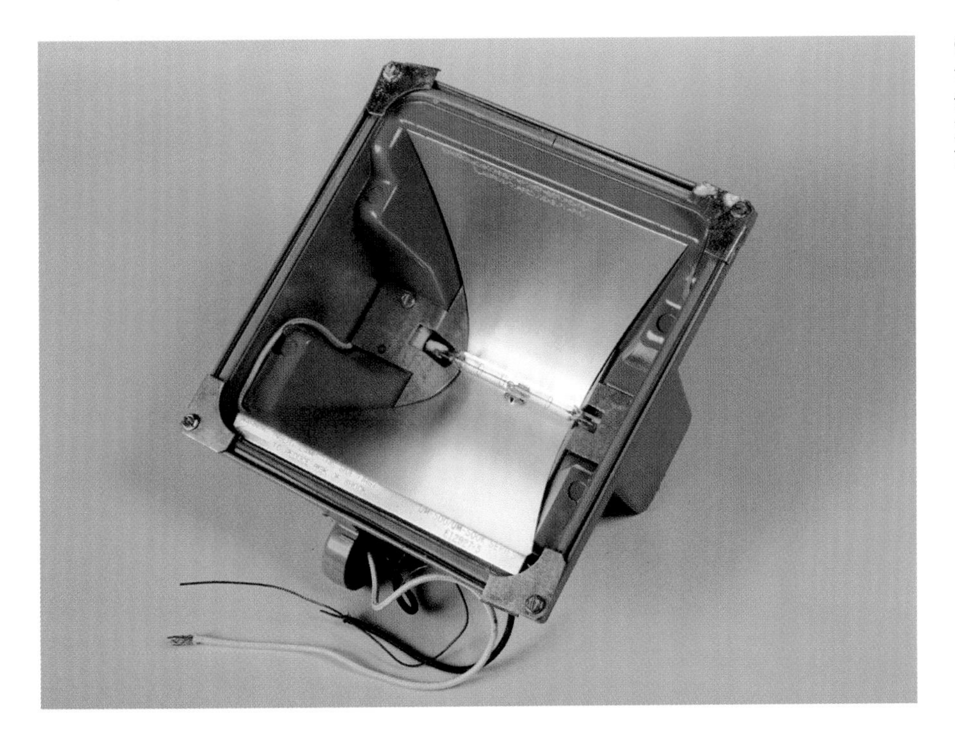

Quartz-halogen floodlights use one-third the energy that incandescent floodlights use and provide better illumination.

the bulbs, which tend to burn out quickly and so need frequent replacement. Also, it's easy to twist the bulb free of its base as you screw it into the fixture socket if you tighten too much (this happened to me just the other day).

When buying floodlight bulbs, be sure to get those rated for the outside, not heat lamps—they look a lot alike.

Quartz halogen

Quartz-halogen floodlights cost more than incandescent floodlights but use around a third of the energy that incandescent floodlights use. Quartz-halogen floodlights are being used by both homes and businesses because of their massive amount of brilliant, natural, color-correct, white light (see the photo above).

The biggest problem with quartz-halogen floodlights is that they give off a lot of heat, so they should not be located where people could touch them. I once left an extension cord lying about 2½ in. in front of a portable quartz-halogen worklight, and the intense heat melted the cord's insulation. It's also a good idea not to drape anything flammable, such as a flag, near the front of the fixture.

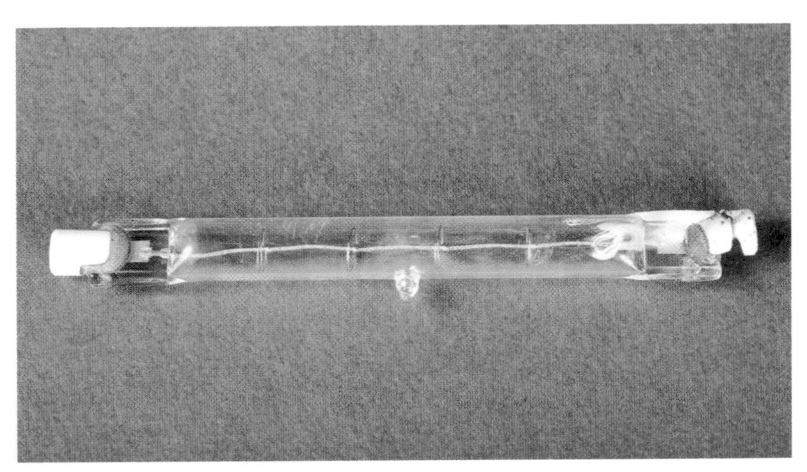

This quartz-halogen bulb blew out on its right side.

For residential units, you can get 300-, 400-, or 500-watt bulbs, and 1,000-watt bulbs are available for large commercial fixtures. The bulbs are supposed to last around 2,000 hours, but my experience has been that they last significantly less than that.

When replacing a quartz-halogen bulb, always be sure the power is off and never touch the new bulb with your fingers. Your fingers will leave oil and moisture on the surface of the bulb, which could cause the bulb to explode when the light is turned on. Wear gloves or use a clean rag, and for added protection, wear safety glasses.

High-intensity discharge

HID floodlights are very popular for area illumination (see the photo below). They work especially well for parking areas and outside security lighting. HID floodlights are more economical to operate than quartz-halogen floodlights because they pull less current while providing six times the illumination.

High-intensity discharge floodlights provide 12 times the illumination of quartz-halogen floodlights and have low operational costs.

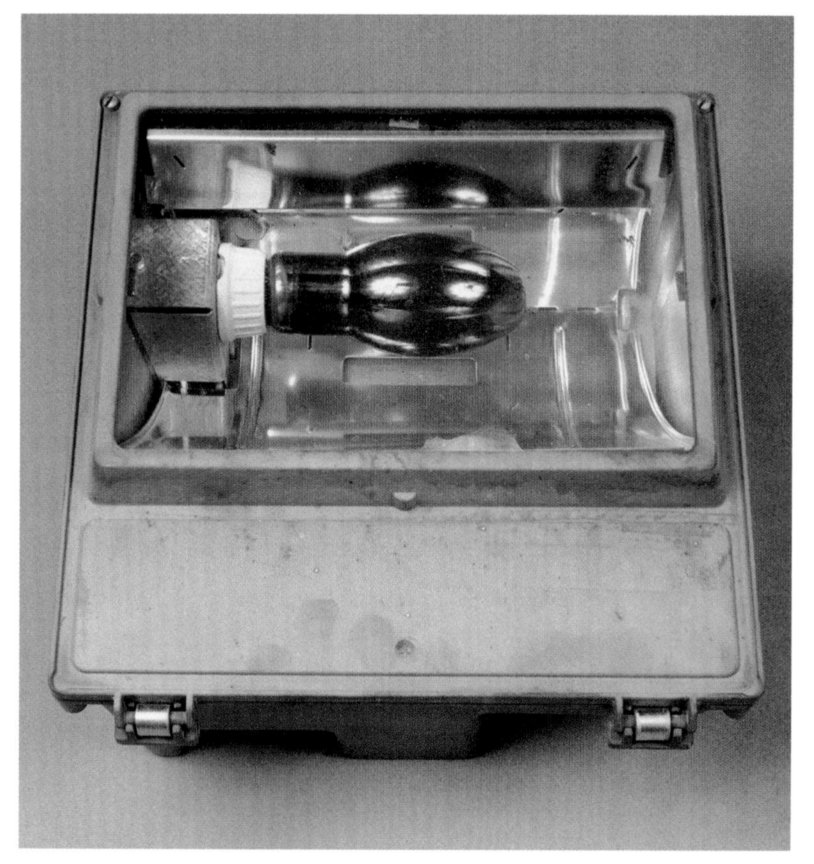

In addition to providing more illumination, HID bulbs last about 12 times longer than quartz-halogen bulbs—around 24,000 hours—and can be screwed into the fixture's sockets.

However, these floodlights do have some drawbacks. One of the biggest is that they have poor color definition—normally yellow-orange—which some people find unattractive and unsuitable for residential lighting. Also, HID floodlights give off a lot of heat, like quartz halogens. They need to be mounted out of reach so that people cannot get burned, and you should not drape anything flammable near the front of the fixture.

And unlike quartz-halogen and incandescent floodlights, which provide instant illumination, HID fixtures need about three to four minutes to warm up.

Installation

Incandescent, quartz-halogen, and HID floodlights are normally 120 volts, and the installation procedures are pretty much the same for each. It's standard practice to mount the fixture on an electrical box—either within the wall or on the outside (see the sidebar on p. 69). But because floodlights are typically mounted high up and are not very visible, you don't have to be picky about getting the fixture to sit flat if you have irregular or beveled siding.

Because a floodlight's beam begins narrow and widens as it moves away from the fixture, it needs to be mounted high to provide the most illumination for large areas, such as driveways, yards, and parking lots. (Try not to illuminate a bedroom window at the same time.) And remember, the higher you mount the light, the higher wattage bulb you'll need to illuminate the area.

The worst part about installing floodlights is that you normally have to be on a ladder, which makes the job difficult, at best. For instance, I recently installed a HID floodlight just below the eave of a two-story house, which tested my abilities as a contortionist. To keep both my hands free, I had to twist a temporary support wire around the fixture base and hang it dangling below its intended location while I locked my legs around the ladder rungs and connected the wires.

Most floodlights have only three wires—black (hot), white (neutral), and a green or bare ground. With only one feeder cable, the wires are simply spliced to like wires in the feeder cable. (If the box is metal, ground it.) Be sure the power is off before you begin working!

If you need to power another light on the same circuit, wire them in parallel. This means there will be an incoming cable and an outgoing cable, and the black, white, and ground wires will be spliced together with pigtails going to the light. It's similar to how a receptacle would be wired in parallel (see p. 39).

Sometimes the power cable will be brought to the light and another cable will be run to the switch. In cases like this, the white neutral wire is spliced directly to the light's neutral. The black feeder wire is spliced to the white wire in the cable to the switch, becoming the hot feeder for the switch (tape it black to indicate it's hot). Then the black wire of that cable is the switched power to the light and is spliced to the black wire on the fixture. All the grounds are spliced together (see the left drawing on p. 44).

However, this should only be done when replacing a light on an existing circuit. In a new circuit, I prefer to take the power cable to the switch box first and then run another cable to the light, so I have easy access to the power when troubleshooting the circuit (see the right drawing on p. 44). And the less wires you have to fiddle with 20 ft. to 30 ft. off the ground, the better.

MOTION DETECTORS

A motion detector is normally installed for security and has a sensor that turns on the lights when it senses heat and movement (see the photo below).

The lighting loads are limited to around 500 watts. Don't exceed this limit, or you could overheat the fixture and the wires. Most motion detectors allow you to control how long the light will remain on once movement has ceased and have a manual override for testing.

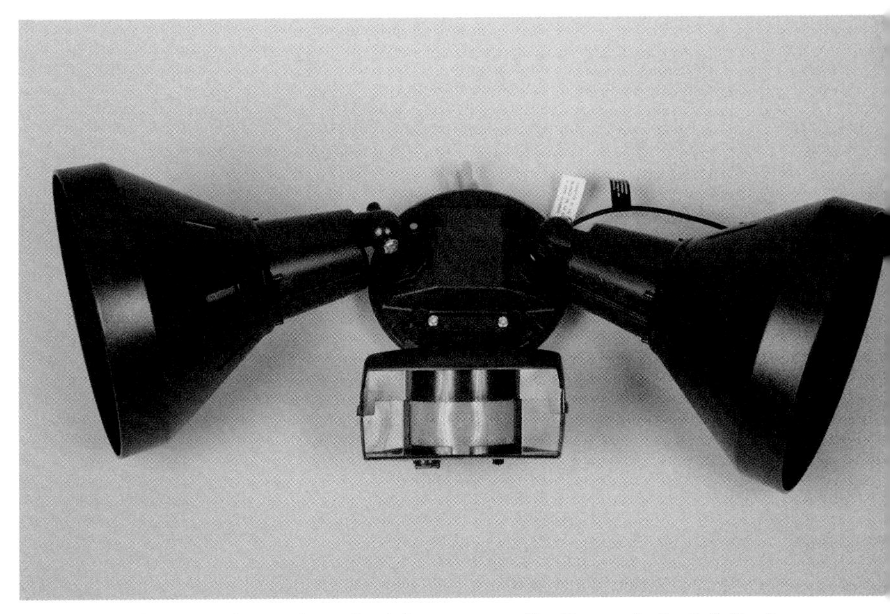

A motion detector is equipped with a sensor (bottom of photo) that turns on the lights when it senses heat and movement.

Motion-Sensor Adapters

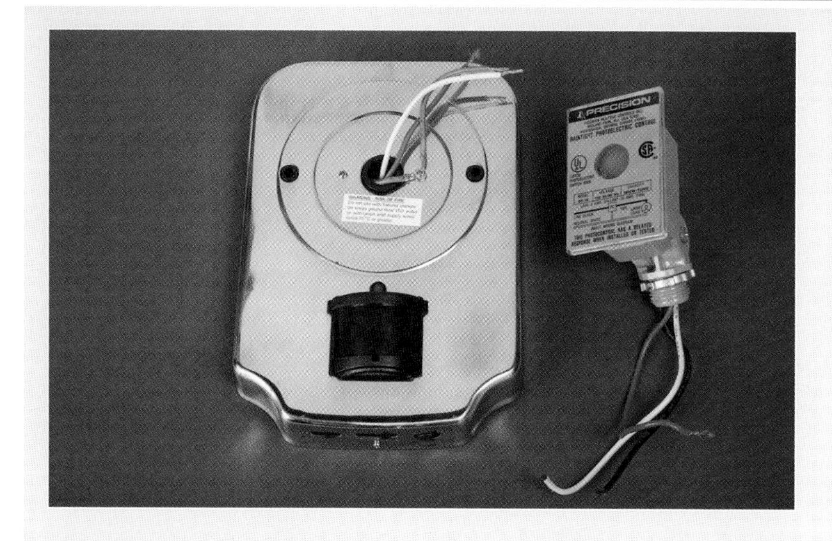

With this type of base (left), almost any light can be converted to a motion detector. Simply remove the light arm from the base of the existing light and attach it to this one. The sensor on the right is attached to a separate light base and can be used to control power to almost any type of light.

You can convert almost any outdoor light fixture to a motion-sensing light with a motion-sensing kit. One version is simply a motion sensor that is cut into the circuit. Another version has the motion sensor mounted in a fixture base.

To install the simple motion sensor (right in photo above), first find a spot for it. Then cut into the feeder cable of the light fixture (you may have to run a separate cable from the sensor to the light's cable). The light fixture will then be controlled by the sensor.

Once you have chosen the location of the sensor, the wiring is pretty simple. The incoming black wire is spliced to the sensor's black wire. The sensor's feed, or output, wire (normally red) is spliced to the light's black wire, and all neutrals (white) are spliced together.

If you bought an adapter kit that has the sensor mounted in the fixture base, here's how to hook it up. First, make sure the light switch is off, then remove the existing light base from the wall, leaving only the mounting box and the feeder cable.

Unscrew the old lamp from the old base and screw it onto the new base. Splice the feeder cable to the wiring of the new base by matching like colors—black to black, white to white, and ground to ground. The sensor output wire in the new base can be almost any color (except white or green), but it's normally red. Splice this wire to the lamp's black wire.

After making the splices, mount the new base to the existing mounting box, and you're finished. Instructions that come with the new unit tell you how to test it. Be sure to turn the power switch back on before doing so.

Before you buy either of these kits, note the wattage limitations of the sensor and make sure it exceeds that of the lamp wattage you want to use. For example, you don't want a sensor that is good for only 700 watts controlling lamps that total 1,000 watts.

You can pay anywhere from $25 to $75 for a motion detector. Better models have adjustable sensitivity of around 15 ft. to 75 ft., and more deluxe models detect movement as close as 3 ft. Be sure you get a unit that has sensitivity controls so that you can adjust for cats, dogs, raccoons, and swaying branches. And make sure the unit has instructions in the box—there are a few models that don't. You'll need instructions to learn how to test, operate, and troubleshoot the unit.

The sensor on the fixture normally samples a 110° view, but some models sample a 360° view.

It's also possible to buy motion-detector adapter kits that can be attached to almost any type of outside light (see the sidebar on the facing page).

Installation

Motion detectors have the dubious distinction of being the most misinstalled of all light fixtures. They are invariably installed too high and/or in the wrong spot to detect motion.

The fixture should be mounted no higher than 12 ft. above the ground (for more on mounting fixtures, see the sidebar on p. 69). The higher the unit is installed, the less sensitive the unit will be to motion. You should also locate the unit away from heat sources, such as a heat pump, and away from reflective surfaces, such as windows or pools, which could cause the unit to malfunction.

Although it may seem logical to face the sensor into the direction of the area to be monitored, it's not the correct way to mount it. A motion detector uses fingers of detection (called lobes) that project from the unit and sense motion when something cuts across them (see the drawings at right). If the unit is installed directly in front of the motion area, it is possible to walk all the way up to the

Locating a Motion Detector

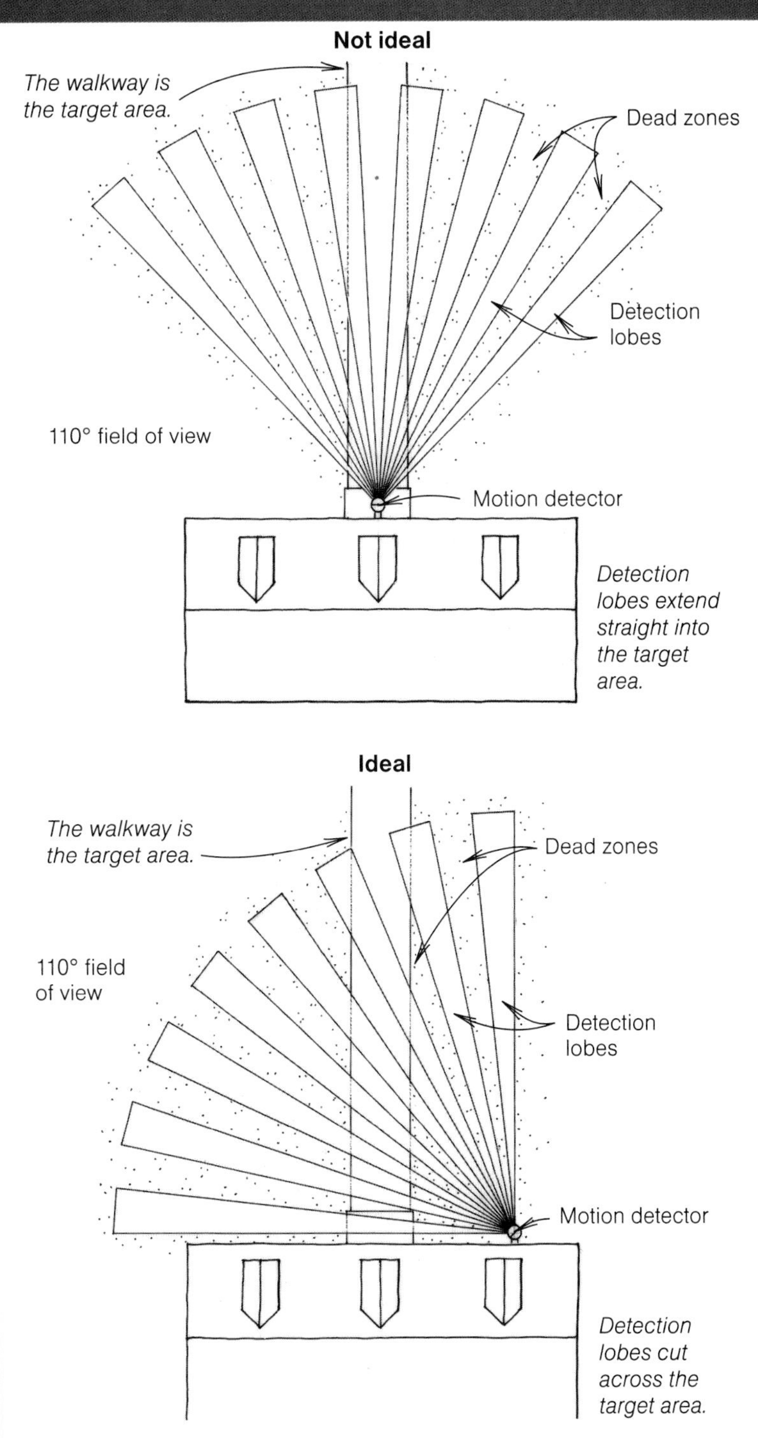

Mount the motion detector no more than 12 ft. above the ground and locate it so that motion cuts across the lobes of detection. Locating the unit straight into the target area may allow someone to walk up to the house undetected through a dead zone.

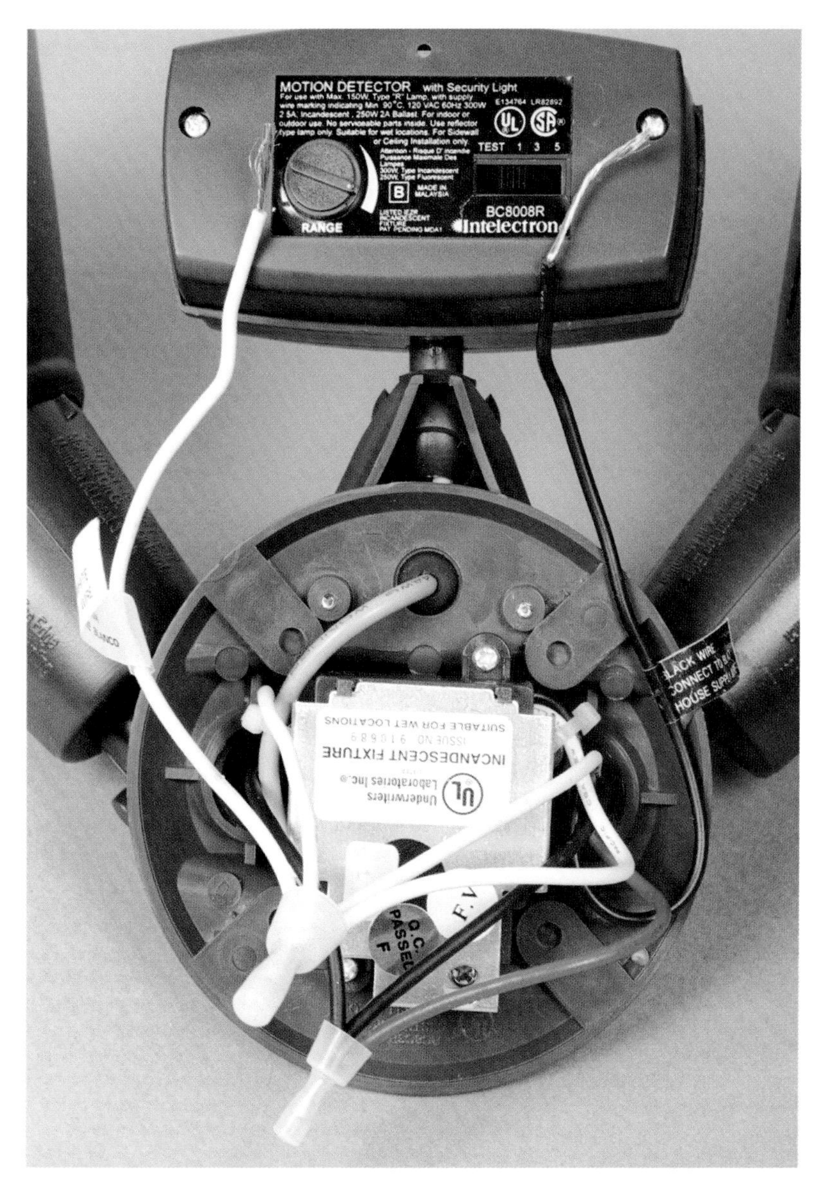

A motion detector is easy to wire: splice black to black and white to white. All the rest is wired at the factory. There is no ground because the unit is plastic.

sensor and not trigger the lights if you walk between the lobes. The fixture must be mounted at right angles to the area you want to monitor so that people must walk through the lobes. If installed for security, you might consider more than one unit at angles to each other to gain more coverage.

Once you have the unit at the correct location, run the cable to it and hook it up. You should wire the light through a switch (for more on wiring switches, see Chapter 3) so that it can be turned on and off for maintenance. The wiring is simple because there are just two wires to hook up: black and white (see the photo at left). All the other wires are internal. Connect them to like wires in the cable (for more on splicing, see p. 40), and you're finished. Most units are now plastic so there won't be a ground. If you have a metal box, attach the ground wire to it; otherwise, simply fold the ground wire back.

After wiring, test the fixture. First turn on the light using the test switch to see if it's wired correctly. Then test the sensor (follow the instructions that came with the unit). Have someone stand in the sensing area and wave his arms to see if it's angled correctly, then have him walk through the lobes at different areas so you can adjust the sensitivity settings and adjust the angle of the sensor if needed.

POST LIGHTS

Post lights are popular decorator items around driveways and sidewalks and can provide safe illumination of these areas for guests. When a post light is equipped with a photo sensor (which turns the light on when darkness falls), this type of light makes an attractive alternative to standard switched outside lighting.

Older models, and some of the heavy-duty ones, will have posts of painted aluminum. But today most residential post lights have plastic posts and plastic light housings because they are less expensive to manufacture and are safer because they are nonconductive. (If you are installing or rewiring a post light that has a metal post, it must be grounded.) In general, how much you pay will ultimately depend on the style of the fixture and on the manufacturer.

Installation

To install a post light, you need to run a cable from the switch inside the house to the outside (see the drawing below). By code, the cable needs to be buried and must be rated for such use: Look for cable that is labeled UF, for underground feed.

Bring the cable out of the house and bury it along its run to the post light (seal around the hole with caulk to keep insects and moisture out of the house). How deep you bury the cable depends on whether the circuit is protected by a ground-fault circuit interrupter (GFCI) or not, although local codes may vary. If it is—and I recommend GFCI protection for safety—most codes say the cable can be buried 6 in. However, to avoid cutting the cable when planting shrubbery, I think it's a good idea to bury it at least 12 in. deep. (If there's an outlet on the post, the

Wiring a Post Light

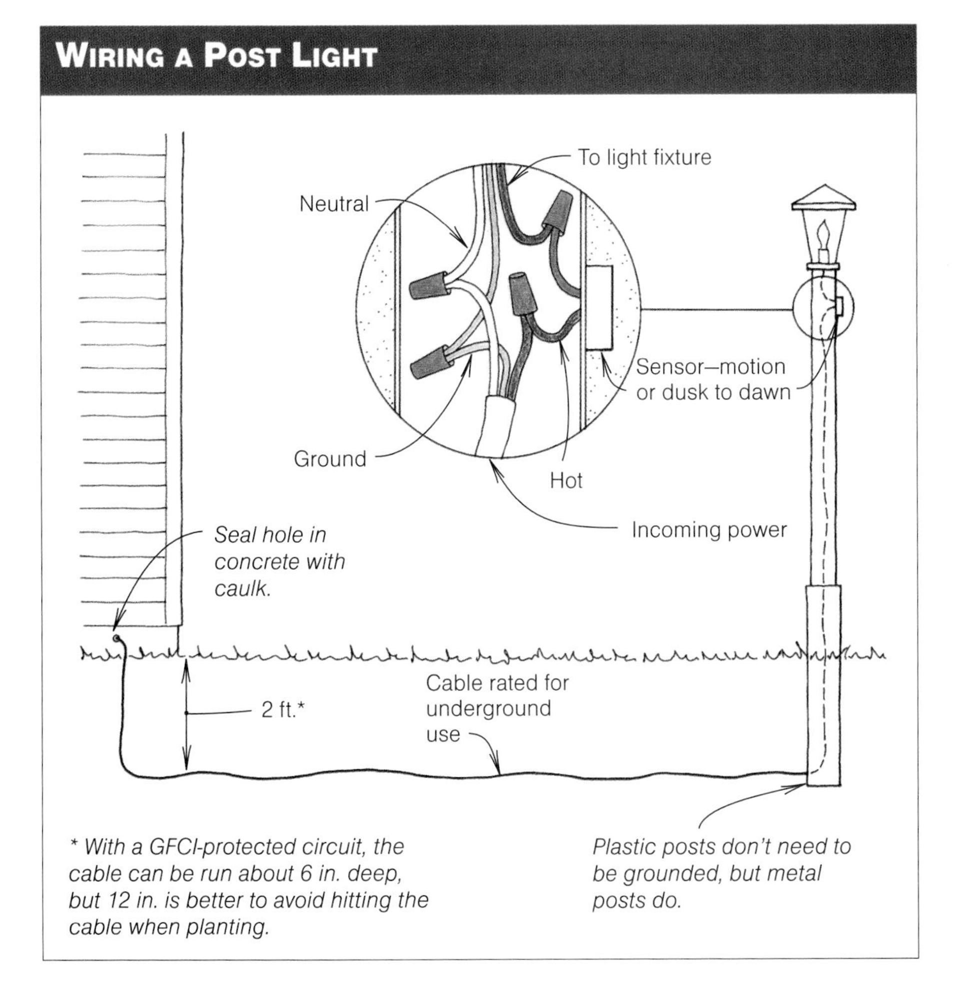

* With a GFCI-protected circuit, the cable can be run about 6 in. deep, but 12 in. is better to avoid hitting the cable when planting.

circuit *must be* protected by a GFCI). If the circuit is not GFCI protected, the cable must be buried 2 ft., and some municipalities require it to be deeper.

Bring the cable to the bottom of the post, which is hollow, and pull it up to the top. Then secure the post in the ground so that it will not tip over. How it is secured is always left to the discretion of the installer, but the post is normally buried around 2 ft. deep, with the earth solidly compacted around it. You can provide extra support with a concrete base, but this is not typically required.

Splicing is done in the hollow post, and making the connections is pretty easy (see the drawing on p. 77). Simply splice all like wires: black to black, white to white, and ground to ground. If there is no ground wire to connect to and the post is metal, ground the metal.

LOW-VOLTAGE LANDSCAPE LIGHTS

Low-voltage landscape lights are a great method of providing accent lighting in a yard, and they are easy to install.

Landscape lights can be purchased at most any home center and come with a transformer (which is plugged or wired into a 120-volt supply), four to six lights, and the wire (which can be 14 to 10 gauge). The kit will cost anywhere from $50 to $100. You can also buy the pieces individually.

The transformer takes the 120 volts and lowers it to a safer 12 volts. Because the voltage is low enough not to harm a person, the wires can be run on the ground or buried in a shallow trench. The transformer has a manual on/off switch to power the low-voltage lights, a photo sensor with an on/off switch, a timer with an on/off switch, or it could have a combination of these.

When buying the kit or the transformer, make sure the transformer can supply enough wattage for the number of lamps you want. Simply add up all the wattage bulbs that will be on the circuit and then buy the transformer that is slightly above that number (at least 10%). If the wattage exceeds that of the transformer, break the system up into two or more independent circuits. For example, instead of having one transformer feeding 20 lamps, have two transformers feed 10 lamps each.

Bulbs can be halogen or incandescent. Most manufacturers sell light kits with low-wattage bulbs—4 to 11 watts—but you can install 75-watt incandescent bulbs for more light. If you plan on using the higher-wattage bulbs, you can't use 14-gauge wire because it could overheat. Instead, use 10- or 12-gauge wires (for more on wire gauge, see p. 15).

There are many styles of landscape lights available, and which you choose will depend on what is to be illuminated and what type of lighting effects you are looking for. First decide what must be illuminated. Is it a driveway, a sidewalk, steps, a patio, trees, ground cover, a pond, shrubbery, a sign, a statue, or even a fence? Once you've determined what is to be illuminated, next decide on the lighting effect (see the drawings on the facing page). Some examples are silhouette lighting, shadowing, spotlighting, spread lighting, path lighting, security lighting, and traffic lighting.

For accent, spot, or spread lighting, the goal is to highlight the desired object without overlighting it. For path, traffic,

Outdoor Lighting Effects

Silhouetting

Locate the light behind the object to be silhouetted and close to it. To create the effect, shine the light onto a vertical surface.

Shadowing

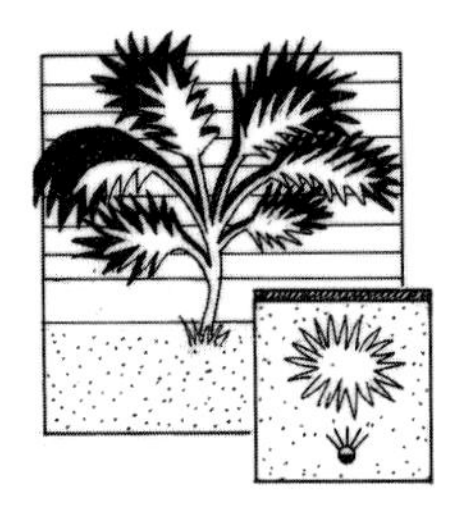

Place the light in front of the object to be shadowed. Shine the light on the object to create a shadow on a vertical surface.

Spotlighting

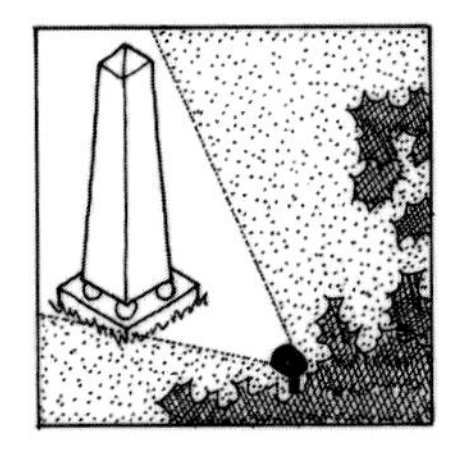

Draw attention to a specific item in the yard, such as a statue, by beaming the light directly onto it.

Spread lighting

Set the light low near ground covers or low shrubbery to create symmetrical patterns of light to highlight their shapes, colors, or textures.

Path lighting

Set the light low to create symmetrical patterns of light to illuminate pathways, sidewalks, borders, and steps. The effect is that of an airport runway.

Security lighting

Security lighting is best created by combining low-level lighting, such as traffic and path lighting, and high-level lighting, such as spotlighting. The result is a well-lit yard with lights that are not harsh or glaring.

Traffic lighting

Traffic lighting combines spread and path lighting. Set the lights higher and group them to illuminate entire walkways, patios, or gardens.

and security lighting, the object is to provide continuous lighting or pools of light over an area. Once you are sure of the lighting effect you want, buy a lighting kit that will achieve the desired result and that will blend into your landscape.

Installation

Installing low-voltage landscape lights is quite simple. In most cases, the lights themselves are mounted on stakes that are driven into the ground. Then it's a simple matter of plugging in the transformer to a receptacle.

If the transformer is plugged into an outdoor receptacle, the outlet must be covered with a watertight, while-in-use cover. This type of cover looks like a large bubble around the receptacle and allows the transformer to be plugged into the outlet while the cover is closed (you can get them at hardware stores). Such a cover keeps rain away from both the transformer and the receptacle (which must be GFCI protected). If the transformer is so large that it prevents the cover from closing completely, it must be kept inside (but it still requires GFCI protection).

It's very important that you minimize the voltage drop (the amount of voltage lost along a conductor from the power source to the load) along the wire run. The voltage drop will decrease the amount of illumination delivered by the lights. For instance, if the lights are supposed to receive 12 volts for full illumination, and they are only receiving 10 volts (a drop of 2 volts), the lights will operate at only partial illumination. It is advisable to keep the voltage drop to less than 2 volts. If you are using halogen lamps, keep the voltage drop to below 1.2 volts. To minimize the voltage drop, plan for shorter cable runs or use heavier-gauge wire to deliver more voltage.

Here's how to calculate voltage drop:

1. Add up all lamp wattages.

2. Multiply the total wattage by the length of wire and then divide that number by the cable constant to find the voltage drop. (The cable constant is a number given by the manufacturer that incorporates the resistance of the wiring.) The cable constants for 14-, 12-, and 10-gauge wire are 3,500, 7,500, and 11,920, respectively.

For example, say you have a 100-ft. length of 12-gauge wire powering five 27-watt lights. Plug the numbers into the formula:

$$(5 \times 27) \times 100/7{,}500 = 1.8 \text{ volts.}$$

That means the voltage drop is 1.8, and the lights will receive 10.2 volts, which is within the recommended range but not ideal. To lessen the drop, use heavier-gauge wire or shorten the run of wire.

TROUBLESHOOTING

Troubleshooting exterior light fixtures follows the same logic as troubleshooting interior fixtures. Often, if your fixture came with instructions (unfortunately, most don't), it will provide troubleshooting directions for that specific light. But here are a few general guidelines.

If the lights just don't work, check the bulb first. Remove the existing bulb and replace it with a known good bulb (for more on troubleshooting light fixtures, see Chapter 4). If the bulb works, the existing bulb is bad.

If the problem is not with the bulb, check the center contact in the socket to be sure it is not pushed down too far or covered with dirt or corrosion. You can bend the contact back up about ⅛ in., but don't bend it too far or it could break. If you see dirt on the contact, clean it with a pencil eraser. Make sure power is off for either of these operations.

If the light is wired through a motion sensor separate from the lights, check the wiring to it and check to see if the settings are correct (refer to the instructions that came with the unit to find specifics about troubleshooting). These sensors can go bad and may need to be replaced. If the motion sensor is part of the fixture, check the settings and try replacing the sensor.

Before buying a new sensor, though, first check that the switch is working properly. Simply measure the voltage across its two terminals with a multimeter (for more on troubleshooting switches, see Chapter 3). This can be done with the switch still in the box and the power on (but be very careful). Remove the cover plate and place one probe on one screw and the second probe on the other. To avoid shorting out the probes on the sides of a metal box, add electrical tape to the metal part of the probes, except for the tip. The voltage should be around 120 volts when the switch is off and close to zero when the switch is on.

You can also check the switch while power is off, but the multimeter will have to function as a continuity tester. Take the switch out of the box and remove one of the two wires on the switch's screw terminals before testing to eliminate any false readings. The switch should have continuity (a closed circuit) across its two terminals when the switch is in the on position and no continuity (an open circuit) when the switch is in the off position.

If you know the switch is working, the next step is to check the fixture's wiring and verify that it has power. To do so, remove the fixture from its box to get to the wire splice. Temporarily hang the fixture adjacent to the box with a piece of wire. Look at the splices and note if they are loose or tight. Remove the wire nuts and look closer. With the switch on, measure the voltage from hot to neutral (the meter should read 120 volts across the black and white wires). If the voltage is 120 volts, the problem is not with the wires to the fixture. It is with the fixture, which probably needs replacement.

If you have a post light that's not working, first check to see if a breaker or the GFCI has tripped. If neither has, check the bulb. If the bulb is good, remove the light head from the post and use a multimeter to verify that it is receiving power.

If the lamp is not receiving power, and you have checked the breaker and GFCI, the problem most likely lies with the light sensor on the post. This type of sensor does go bad. To test the sensor, you must first verify that it has power—120 volts across the incoming black and white wires. Under normal daylight conditions, the sensor will have power going to it and no power coming out of it. To see if the sensor works, cover the sensor head with a couple of layers of black electrical tape. Wait a few minutes for the photocell to sense the loss of light and then measure the output voltage—normally across a red wire and the white wire. No power means a bad photocell. If you do measure voltage, the wiring or the socket is bad and will need to be replaced.

6

CEILING FANS

A couple years ago I was called out to a home to repair a ceiling fan gone wild. When the fan was running, the homeowner said, it wobbled so violently that the blades were hitting the ceiling. "Yeah, right. Hitting the ceiling," I thought. "Probably just a ploy to get me out to the house faster."

When I arrived at the house and turned on the fan to see the problem for myself, my eyes bulged and my jaw dropped. I was looking up at the ceiling fan in action. It was installed on a vaulted ceiling and was swinging so far that it was chopping chunks out of the ceiling. I turned off the fan immediately, but not before small pieces of the ceiling hit the floor.

I immediately knew what the problem was, and it wasn't with the wiring. The fan was slightly unbalanced, and the drop pipe (the supporting rod from the ceiling to the fan) was too short, allowing the blades to hit the vaulted ceiling.

CHOOSING A FAN

When shopping for a fan, there's a lot more to think about other than how much it costs and how it looks. I'm not saying to ignore these two criteria. I'm just suggesting to add a few more. You must also consider the warranty of the unit, the space of the room it will be located in, the blades, the motor, and the accessories that come with it or that can be added.

Price and looks

There are a lot of cheap ceiling fans out there, but remember, you get what you pay for. That's why I recommend paying a little more for a ceiling fan to get one that will last, instead of going for the cheapest price and having to replace the fan a few years down the road.

Be sure you like the looks of the fan, because it will be a prominent addition to a room. You may want the fan to match the decor, or you may want something different. For instance, when my cheap living-room fans went bad (yep, I made the mistake of going for the lowest price), I wanted to replace them with something that didn't look like an ordinary pancake-style fan (one with a flat housing near the ceiling).

I wanted something old-fashioned and more ornate, and that's what I bought (see the photo on p. 84). Now, instead of just walking under the fans in the room, people gaze up, admiring them. There are many ceiling-fan designs out there, and you should have no problem finding one to suit your tastes.

Summer Cooling, Winter Warming

A ceiling fan doesn't have to be put to rest during the winter months. A reversible ceiling fan not only provides cool breezes in the summer, but it also helps move warm air in the winter.

In the summer the fan cools by pulling air up along the cool exterior walls, bringing it to the center of the room, and then forcing it down again (see the top drawing below). The cooling effect provided by this downward-flowing air is large enough to allow the air-conditioning thermostat to be raised by 6° to 8°, cutting air-conditioning costs by as much as 40%.

In the winter, a ceiling fan can help keep you warm by moving warm air trapped at the ceiling down to living areas (see the bottom drawing below). The fan can be switched to its lowest speed setting and its rotation reversed to pull air up and then push down the warm air. Running the fan in this manner will distribute heat better, which can save 5% to 10% on heating costs.

Summer cooling

The fan cools by pulling air up along the exterior walls, bringing it to the center of the room, and then forcing it down.

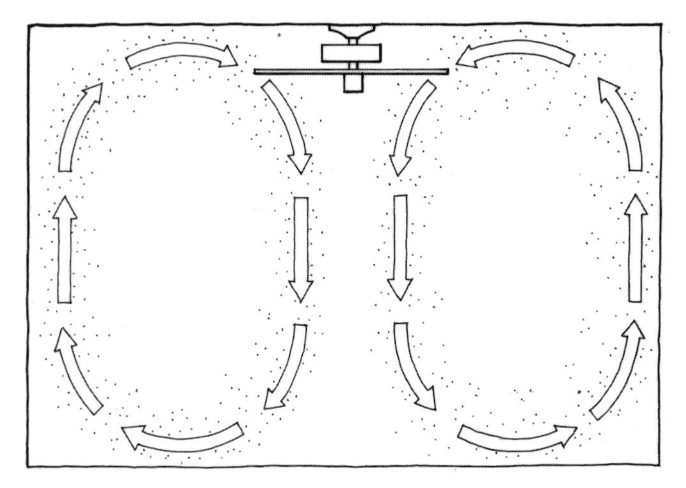

Winter warming

With the fan in reverse mode and set on low speed, warm air trapped at the ceiling is circulated downward along the exterior walls.

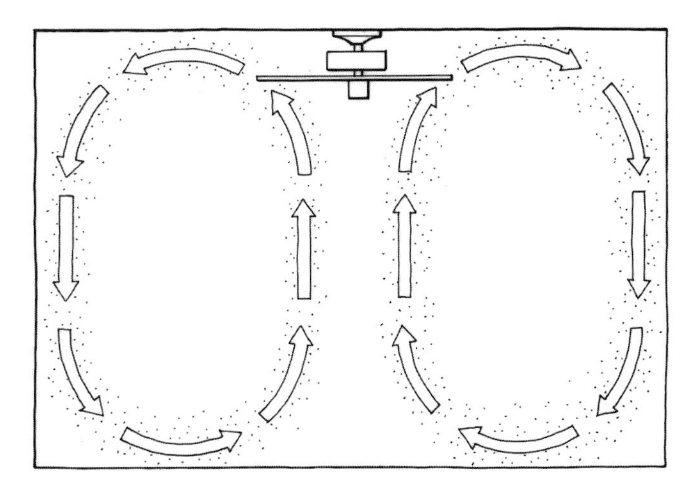

You don't have to stick with plain ceiling-fan designs. There are a lot of options available, like this replica of a unit from the late 1800s made by Hunter.

Warranty
Be sure that the fan you choose has a good warranty. A 20-year to lifetime warranty is very good. If the company doesn't offer a long warranty, there might be a reason why.

Space requirements
When choosing a ceiling fan, you also must consider both the size of the room and the size of the fan to be installed. Be sure the fan is large enough to move air in the room it will be located in (see the drawings on the facing page). And be certain you have enough headroom in its proposed location. The blades need to be a minimum of 7 ft. from the floor and 18 in. from walls. You don't want to knock people in the head as they walk under the blades, and you don't want the blades to hit a wall.

Also, almost all fans come with a three-speed reversible motor (see the sidebar on p. 83). The speeds are adjusted using a pull chain, and the reversing switch is normally mounted on the side of the housing. If the unit is to be mounted on a high ceiling, you won't be able to reach these controls easily. In a situation like this, you may be better off choosing a unit with a remote control or wall-mounted controls (see p. 86).

Blades
The goal of buying a fan is, of course, to move air, and I don't recommend buying a fan that doesn't list how much air it can move. Air movement is measured in cubic feet per minute (cfm). The higher the cfm, the more air the fan can move.

The blades and motor combine to move air. When thinking about the blades, consider what they are made of, their weight, their size, and their pitch. Blades should be solid wood and sealed to prevent moisture from warping the wood. (If the blade warps, the fan will wobble.) Most manufacturers put a different finish on each side of the blade, allowing you to flip the blades to alternate colors and finishes or to keep them the same.

Just as car tires need to be balanced for smooth operation, fan blades need balance as well. That means the blades should be close in weight (within 1 gram is a good tolerance). If the blades are not close in weight, the fixture will wobble. You can get small weights from the manufacturer to add to blades to balance them (I know some people who have taped pennies on the blades to balance them). But it's better to buy a unit with quality blades that are balanced from the start.

Be sure the blades are large enough to move air efficiently within the room (see the drawings at right). Fans have diameters ranging from 32 in. to 52 in. For a room with an 8-ft. ceiling, follow these guidelines: Choose a 52-in. fan for rooms up to 400 sq. ft.; a 44-in. fan for rooms up to 225 sq. ft.; a 42-in. fan for rooms up to 144 sq. ft.; and a 32-in. fan for rooms up to 64 sq. ft. Larger rooms and rooms with higher ceilings may require more than one fan.

Also be sure the blades have sufficient pitch (angle of the blade) to move air. Without sufficient pitch, the blades just turn without moving air. For maximum air movement, I recommend buying a unit with a blade pitch of 15° to 16°. Thirteen degrees is a more conservative pitch, with 11° being typical for a cheaper, poor-quality ceiling fan. A good 52-in. blade with a 15° pitch can move air at a whopping rate of 6,900 cfm.

Motor

The motor is the heart of the fan, and as I mentioned before, it works together with the blades to move air.

You want a large, powerful motor built for many years of hard, quiet, maintenance-free operation. Look for a motor that is cast iron (see the top photo on p. 86), which will draw heat away from the electrical windings so that they do not overheat. The weight of a cast-iron motor will also help prevent the fan from wobbling.

Be sure the motor has sealed ball bearings (on both sides) that are permanently lubricated. If sealed on only one side, dirt and dust could get into the bearings and destroy them.

Sizing a Fan

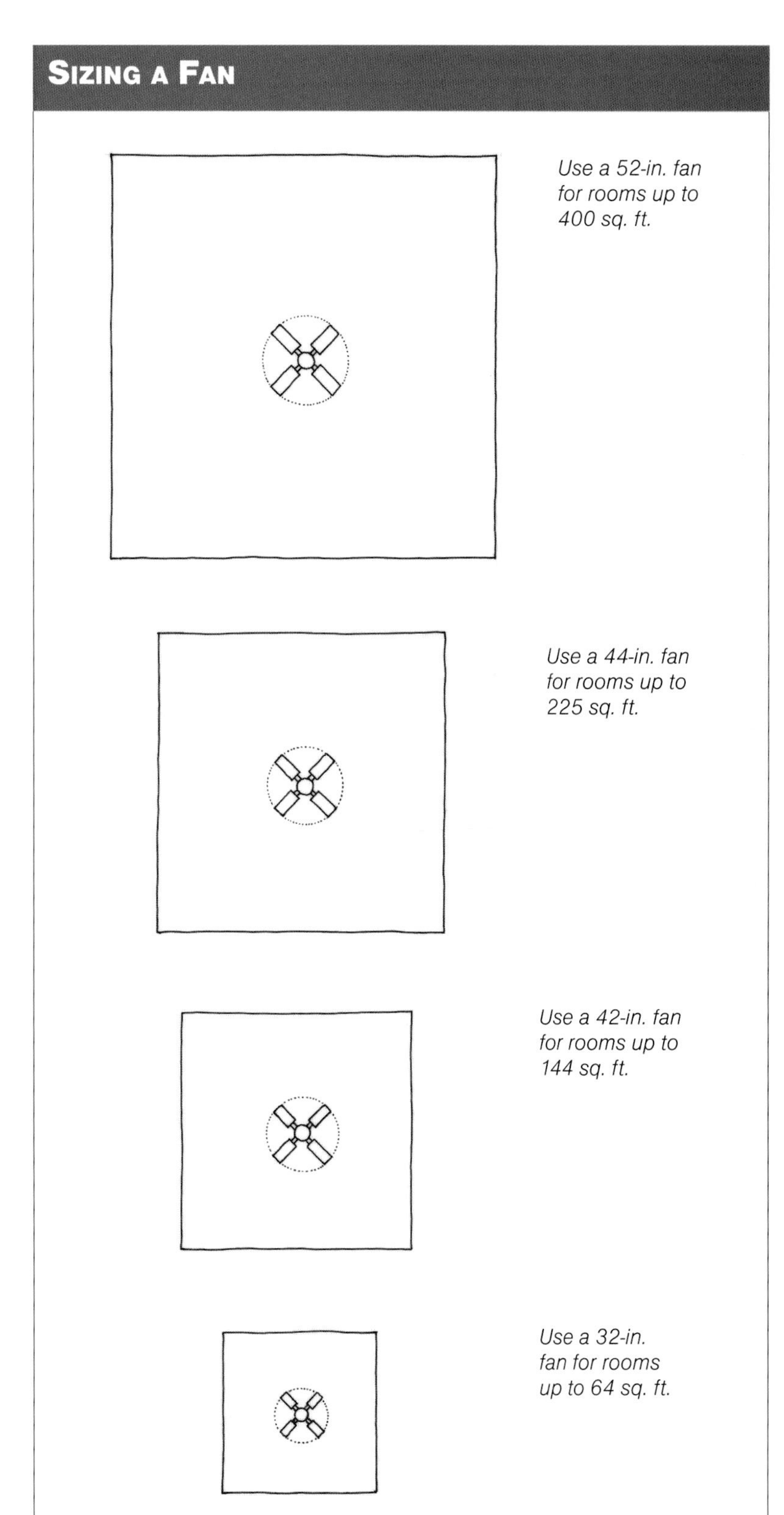

These are guidelines for rooms with 8-ft. ceilings. Larger rooms and rooms with higher ceilings may require more than one fan.

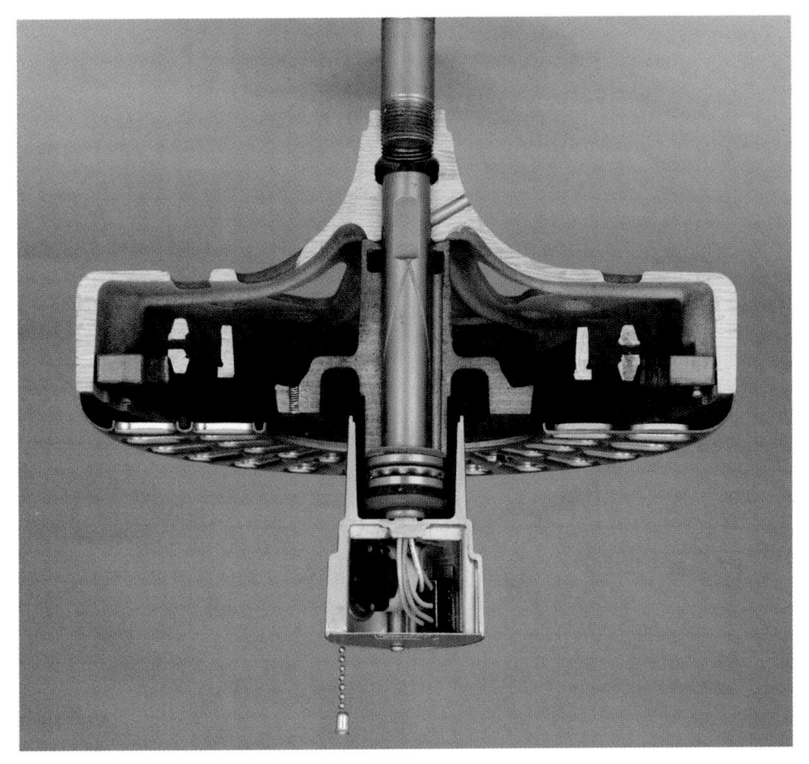

The motor of this fan, made by Hunter, is cast iron, which makes it almost indestructible. The motor also serves as a heat sink and has sealed, lubricated ball bearings to ensure a long life. (Photo courtesy of Hunter Fan.)

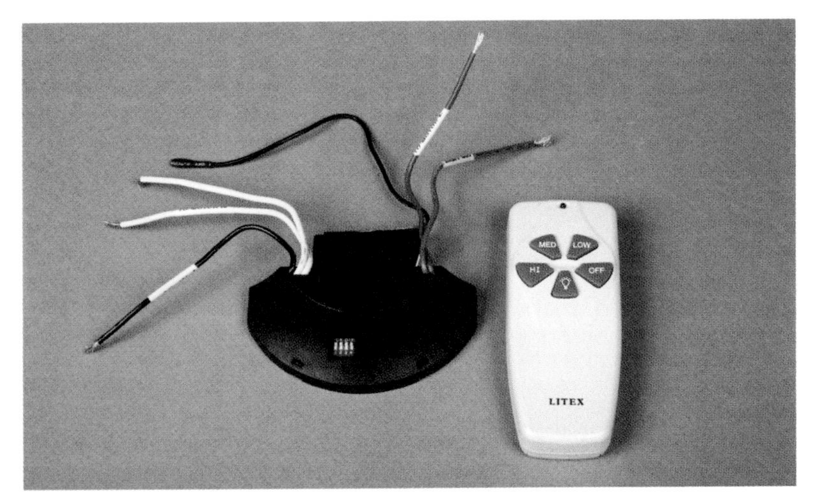

Existing ceiling fans can be wired to a remote control. This unit attaches under the mounting hood. Leave the antenna wire (third from right) sticking out beyond the metal cover to make it easier to pick up the signal; otherwise, the unit may not work well.

When buying a fan, don't let a salesperson talk you into buying one simply based on its speed, or revolutions per minute (rpm). The rpm rating of the fan is not as important as the amount of air it moves (cfm rating).

It is hard to know the quality of the motor before you buy the product (you can't slice open the housing to see the inside). But it pays to ask questions, either of the dealer or manufacturer. Also, make sure the motor is covered by a good warranty. If it isn't, beware.

Accessories

There are two accessories available that can add to the convenience and cost of a ceiling fan: remote-control mechanisms and light kits.

Remote controls are the newest trend in ceiling-fan accessories (see the bottom photo at left). They allow you to turn the fan on and off and adjust its speed and direction from anywhere in the room. A remote control is especially handy for fans on high ceilings.

A remote-control unit is installed under the mounting hood of the ceiling fan (although there are some types that are wired into the wall switch). These units are pretty easy to install (the wires are labeled) and should come with complete instructions. But if the unit has an antenna wire, make sure it sticks out a bit from the housing to make it easier to pick up the signal; otherwise, it may not work well.

Light kits are the most common fan accessory. Almost all fans are made to accept light fixtures and are equipped with all the necessary hardware. My biggest complaints with light kits are that they make the ceiling fan more prone to wobbling (unbalancing them) and that the light fixtures are designed not for function but for form. Cute little teardrop light bulbs don't provide real room-brightening light. However, some manufacturers, such as Hunter, have light kits that can provide up to 300 watts of lights. I find that it's better to go for the higher wattage and then put the light on a dimmer (for more on wiring a dimmer switch, see Chapter 3). This way you can have everything from reading light to night light. The lights can also be controlled by a remote.

INSTALLING A FAN

The most difficult part of installing a ceiling fan is hanging it correctly. I've had countless service calls in which I was called to repair a wobbling fan or to replace a fan that had fallen right out of the ceiling. This is especially embarrassing when it falls on the dining-room table during the evening meal. Common sense and electrical codes need to be followed. Remember the ceiling fan I told you about at the beginning of the chapter? The one that was hitting the vaulted ceiling? Well, I hate to say it, but that installation actually passed inspection somehow. But by following the guidelines here, you should be able to install a ceiling fan so that it will not wobble or fall out. The first thing to do is to securely mount the appropriate box.

The right box

A ceiling fan is a heavy fixture and needs firm support. You can't just hang a ceiling fan from a standard ceiling-fixture box—it will not stay in place. Many homeowners don't realize this. I've been on many a service call to install a ceiling fan where a light used to be, and the homeowner invariably tells me simply to remove the overhead light and just install the fan. They think I'm jacking up my fee when I explain to them that the existing light box isn't made for the weight of an overhead fan and that I have to remove the existing box and install a UL-approved one.

Fan-mounting boxes must be UL listed for such. The box on the left is a pancake box. The one on the right is a standard ceiling box.

Nonmetallic fan-hanging boxes are now on the market. On this unit the spliced wires are stuffed into the side pocket on the right.

Supporting a Heavy Fan

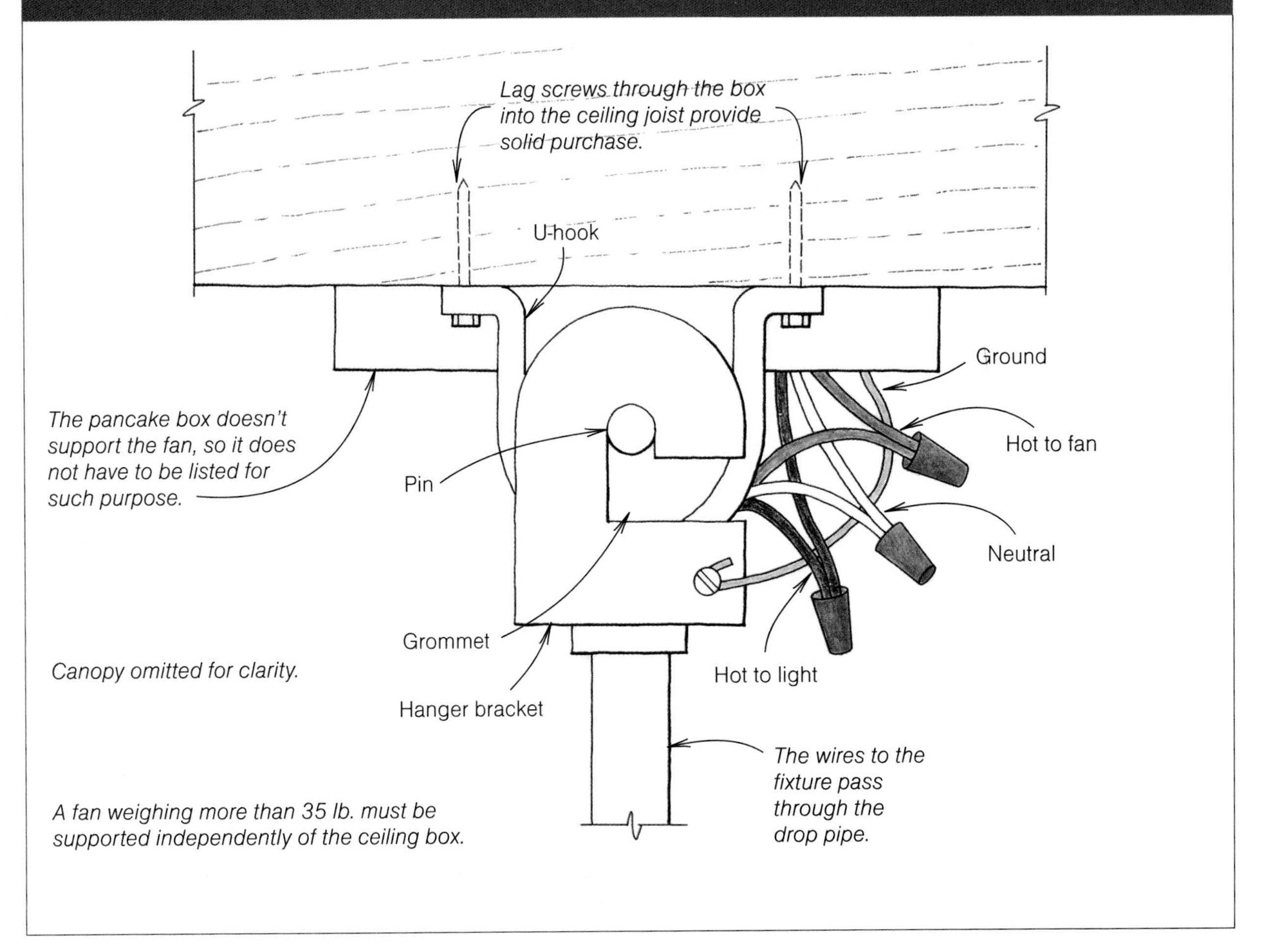

All fan-mounting boxes must be labeled for such (see the photos on p. 87). You can get fan-mounting boxes in both metal and nonmetallic designs.

But if the ceiling fan weighs more than 35 lb., it cannot be mounted onto a box. It must be attached directly to the overhead joists or to another secure mounting surface (see the drawing above). In this instance, because the box is not supporting the fan and is serving as just a splice box, it can be a standard ceiling box.

Mounting

Because of its weight, a ceiling fan must be mounted to a secure wood beam. This is easy enough if the fan is located directly under a joist. But if you want the fan at a certain spot (over the center of the bed, for example), and there is no beam or ceiling joists to attach it to, you'll need to install 2x6 bridging between the joists and attach the fan-mounting box to that (see the drawings on the facing page). Just be sure the front of the box is flush to the finished ceiling.

Three Ways to Support a Fan with Bridging

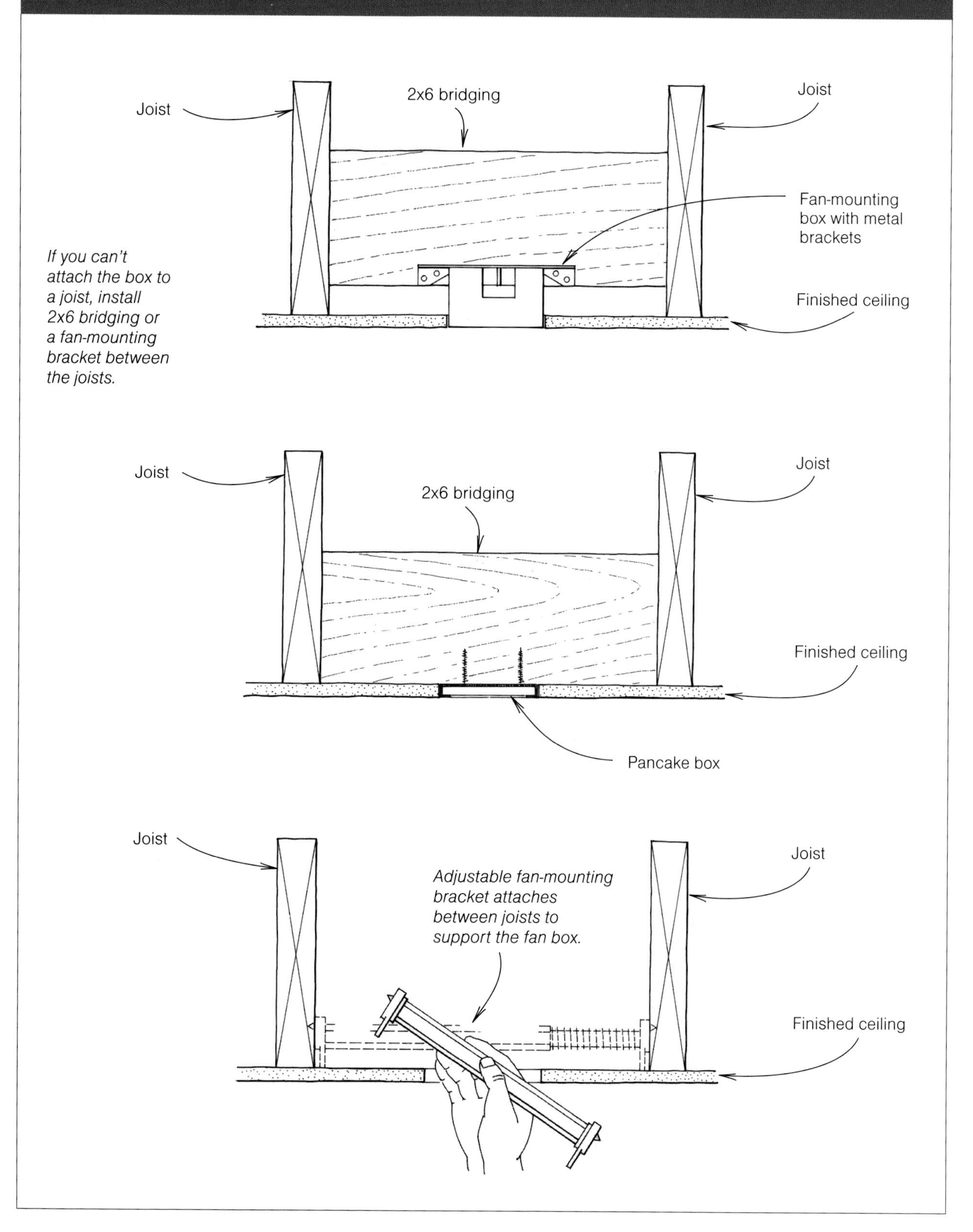

For a room with a low ceiling, the fan must be flush mounted. Unfortunately, this restricts air movement, so it is not ideal.

Installing bridging is pretty simple if the room is not finished, if the ceiling is exposed, or if you can access the spot through the attic. But if you don't have easy access to the joists, you can purchase a special fan-mounting box that has adjustable arms. This type of box is UL-approved for installing ceiling fans, but it can only be used for fans weighing 35 lb. or less.

Installation of this box is easy. First cut a hole in the ceiling for the box itself. Then insert the box into the hole and extend the arms out to catch onto the sides of the joists. Each arm has a wood screw that is driven into the joist by rotating the arm. Having installed several of these myself, I can tell you it's easier to get the arms into the joists at an angle as opposed to straight on. But that's not the right way to do it. If the box is installed at an angle, the screws won't gain full purchase into the joists, so they may not hold, and the fan could drop. Take your time and make sure the arms hit the joists at right angles.

There are three methods of attaching the fan to the ceiling box: flush to the ceiling, hanging with a drop pipe, or angled for a vaulted ceiling.

Attaching the fan flush to the ceiling is most often done in rooms with low clearance. Unfortunately, having the blades so close to the ceiling limits air circulation (see the photo above).

Hanging the fan from a drop pipe increases air circulation. A short (4 in. to 6 in.) ½-in. or ¾-in. drop pipe is included in most fan packages. Longer drop pipes can be used if needed, which may be the case with a vaulted ceiling. The fan wires will have to be run through the center of the drop pipe and should be spliced at the overhead splice box (as shown in the drawing on p. 88). Do not splice in the pipe.

A ceiling fan installed in a vaulted ceiling sometimes needs a special housing to allow it to hang straight down (see the photo below). Make sure the drop pipe is long enough to allow the blades to spin without hitting the ceiling.

If you are installing a fan in a vaulted ceiling, you may need a special attachment housing that fits against the angled ceiling, allowing the pipe to hang straight down (see the photos on this page). Many ceiling-fan kits are universal, allowing both straight and vaulted installations as long as the vault is not too severe.

Wiring

Ceiling fans can work off both 14- and 12-gauge wires (for more on running wires, see Chapter 3). Before wiring the unit, make sure the power is removed from the circuit. The fan will come with four wires ready to be attached: black, red (this color may vary with the manufacturer), white, and ground. Simply splice like wires and make sure the splices are tight (for more on splicing, see p. 40).

If you are going to install a light with the fan and control the power with a wall-mounted switch, use three-conductor cable (three-way switch wiring) from the

Inside the housing, a ball socket fits into the hanging bracket. The bracket is installed on an angle, but the drop pipe hangs straight down.

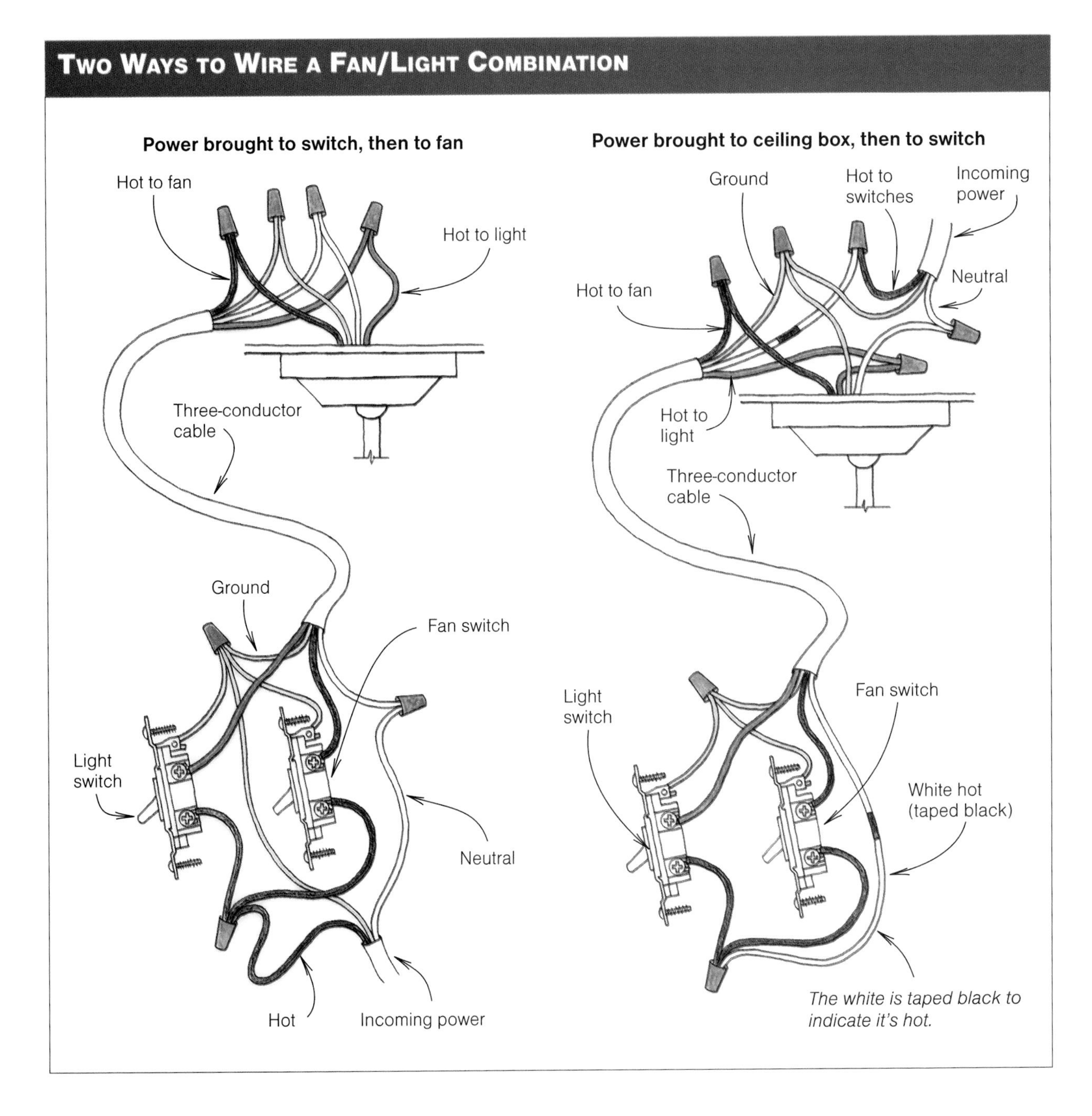
Two Ways to Wire a Fan/Light Combination
Power brought to switch, then to fan
Hot to fan
Hot to light
Three-conductor cable
Ground
Fan switch
Light switch
Neutral
Hot
Incoming power
Power brought to ceiling box, then to switch
Ground
Hot to switches
Incoming power
Neutral
Hot to fan
Hot to light
Three-conductor cable
Light switch
Fan switch
White hot (taped black)
The white is taped black to indicate it's hot.

wall switch to the light. (Even if you aren't going to install a light with the fan at this time, it is a good idea to go ahead and run the extra wire just in case you change your mind later.) This will allow you to install a switch for the fan and one for the light. Black is normally the hot wire for the fan motor; a red wire (or other color) will be the hot for the light; a white wire will be the neutral for both; and a bare or green wire will be the ground (see the drawings on the facing page). Although the drawings show two ways to bring power to the fan, it's best to bring the power cable into the wall switch rather than into the fan outlet box. This simplifies the overhead wiring and allows easy access to the power cable for troubleshooting.

If the fan/light combination shares the same hot conductor, you cannot install a dimmer switch or a fan-motor switch on the wall. The dimmer will affect the motor control, and the fan-motor switch will control the light as well. A switch that does both is not available, so you'll have to install a standard switch and control the fan and light separately at the fixture.

If you choose to use a remote control for the fan, you don't have to install a wall switch—just bring power to the outlet box in the ceiling. This, of course, is perfect for renovation where the walls are already up, and it would be very expensive and difficult to run the switch wiring.

Once the wiring has been completed, attach the cover plate and the fixture canopy, and test the fan.

TROUBLESHOOTING

Most troubles with a ceiling fan are with the installation, not with the wiring. But if your fan does not work, and its pull chain is on, first make sure the breaker is on, then use a multimeter to check the switch on the wall. Measure across the hot wire and the neutral: It should read 120 volts (or close to that).

If the switch is okay, measure the voltage at the ceiling box. It also should read 120 volts (see the photo at right). If it doesn't, the switch or wiring is bad. Make sure all the splices are still tight. The very last thing to do is check the continuity of the pull-chain switch (follow the same guidelines for a regular switch—see p. 49). If all of these items check out, it probably means that the ceiling fan is dead and will need to be replaced.

If the fan doesn't work and the switch wiring is okay, verify that the fan has voltage. Remove the wire nuts and touch the red lead to the black wire and the black lead to the white. The meter should read 120 volts, or close to it.

Although the wiring can sometimes cause a ceiling fan to malfunction, the two most common problems are wobble and humming. If the fan is wobbling, first make sure that it is mounted 100% secure and that there is no wiggle at all in the outlet box or its supporting joist. If you can wiggle the fixture, you should remove it and install it more securely to a ceiling joist or to bridging.

If the fan still wobbles after remounting or reinforcing it, try balancing the blades by adding weights to them (see p. 84). I think the best solution is to replace the blades with new ones. However, some fans will wobble no matter what you do.

Humming is another common problem with ceiling fans, but it's hard to fix because the source lies with the fan's design. A fan that hums is typically a poorly designed, cheap unit, where the manufacturer skimped on materials. However, humming sometimes comes from the fan's variable-speed control. If you can, replace it with a better control made by the manufacturer just for that particular fan. Some controls are now labeled as quiet.

7 BATHROOM FANS

I once received a call from a homeowner complaining that his bathroom fan was not working. I responded within a few hours, but I was still too late to save the fan—and the bathroom. The bathroom was riddled with moisture damage from the humid air, so I could tell that the fan had not been working properly for several months.

As this homeowner witnessed, moisture-laden air can cause a lot of damage in a bathroom over time, such as corroded light fixtures, rotted wood, and peeling wallpaper and paint. It can also hasten the spread of mold and mildew. A bathroom fan is supposed to remove that moisture-ladened air from within the bathroom and pull in dry air from outside the room.

To prevent moisture damage, most building codes require fans in bathrooms that have no windows. But I think a fan is a good idea in a bathroom with windows, too, especially in cold climates (it's silly to think a person will open the window to ventilate the room when it's freezing outside). The more moisture you remove from the bathroom, the better off you are.

CHOOSING A FAN

Many folks who have bathroom fans already in place don't use them, either because the fan is too loud (I call these ring-a-ding fans) or because it doesn't move enough air. These are common complaints, but the problems can be avoided before the installation by buying a good-quality, quiet fan. There are many types of bathroom fans out there, and the choice may seem simple: choosing the unit with the best price. But remember, the best price may not be the cheapest price—you get what you pay for.

Whey buying a bathroom fan, keep these three things in mind: First, choose the right size fan for the room it will be installed in; second, it pays to buy a quiet unit, which will mean paying a bit more; and third, you should also consider some of the accessories available that can make the unit more versatile (or problematic).

Size

First consider the size of the bathroom itself and then choose a fan large enough to move air adequately for that size room. You don't want to put a tiny fan in a large bathroom because it won't move enough air.

The volume of air the fan moves is measured in cubic feet per minute (cfm) and will be listed on the unit. To choose the fan with the correct cfm rating for the room size, first figure the volume of air in the bathroom and then divide that number by 7.5 as per the Home Ventilation Institute guidelines.

Typically you would choose a 50-cfm unit for a small bathroom (around 5 ft. by 8 ft.), an 80-cfm unit for a midsize bathroom (around 8 ft. by 8 ft.), and a 110-cfm unit for a large bathroom (around 10 ft. by 10 ft.). (These figures are based on the bathroom having an 8-ft. ceiling.) For bathrooms larger than 10 ft. by 10 ft., Panasonic makes a 190-cfm unit.

In general, it's better to move too much air than too little. If you are caught between two sizes that are close to your calculations, always choose the unit with the higher cfm rating. In all likelihood, the price difference between the two will not be that great.

Quiet, please

Bathroom fans have a reputation for being noisy. And unfortunately, noisy fans are installed more often than quiet fans because they're cheaper and are usually the only ones stocked by home centers and suppliers, so they're the contractor's choice. I called a major plumbing supplier that sells several million dollars of material every year to see if the company had any quiet fans in stock. They didn't. (They do now because of my complaints.)

But don't be fooled into thinking that your only choice is a noisy unit. Most manufacturers *do* make fans that are quiet and built to last—you just have to shop around and ask for them (or make a special order). It will take a longer time to get the fan, and it will cost more, but believe me, the quality will be worth the money and the wait.

In general, the higher the cfm rating of a fan, the more noise it makes. But the noise doesn't have to shake the floor. You should choose the quietest unit for the amount of air that must be moved.

A fan's noise is measured in sones (see the Glossary on p. 146), and each unit should have this measurement clearly labeled.

A noisy fan is one that is above 3 sones. You'll be able to hear a fan this loud even while the shower is running (it will sound like a helicopter is landing on your roof). A quiet fan, on the other hand, measures around 2 to 3 sones. An ultraquiet fan will be 1 sone or less. With this type of fan, you'll only hear a soothing whooshing noise as the air moves through the unit.

But beware of false advertising. I've seen some manufacturers advertise their fans as being ultraquiet but still have a noise level above 3 sones. And I've seen some fans advertised as quiet that were above 10 sones. So don't listen to the manufacturer's noisy advertisement. Check the label on the fan's box, or take the fan out

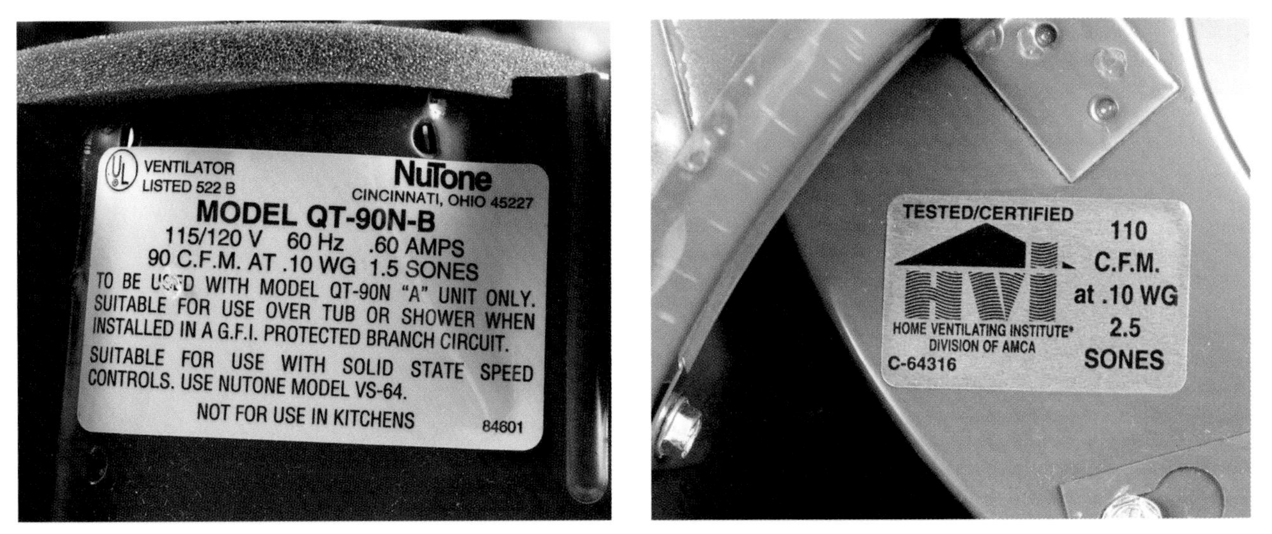

In general, the higher the cfm rating of the fan, the louder it will be. The fan in the photo at left is rated to move 90 cfm of air and is considered quiet at 1.5 sones. The unit on the right moves more air (110 cfm) but is noisier at 2.5 sones.

of the box and check the sticker inside the unit (see the photos above) or the spec sheet in the box.

Combination units

It's very common for manufacturers to sell fan units that are combined with lights or heaters or both. But these units come with inherent problems.

Fan/light combination units do not provide great illumination, especially if they are limited to 60-watt bulbs. But the biggest problem with fan/light combination units is that most manufacturers make the covers for the lights out of cheap plastic. Over time the cover turns brown and degrades as a result of constant exposure to the ultraviolet (UV) rays and heat from the light.

For both of these reasons, I suggest keeping the light separate from the fan: Put the fan in the best location to get rid of the moist air (as close as possible to the tub or shower stall), and put the light in the best location to illuminate the bathroom (in the middle of the room or near the sink).

If you forego my advice and decide on the fan/light combination unit anyway, choose one that has a glass cover over the light. You may pay more for the unit, but it will look better a few years down the road. You can also buy a unit with a plastic cover that is UV and heat resistant so that discoloration won't be a problem, but this is a relatively new addition, and you may have to search a bit to find this type of unit.

Even though fan/light/heater combination units are popular—and I install them all the time for clients—I don't really like them. To me, it makes no sense to put the heater in the overhead fan because you are just heating the air and then blowing it outside. I think it makes more sense to install a separate heater in the wall or in the ceiling near where you step out of the shower or tub.

Also, sometimes the heat from the infrared lamps gets too intense, making you feel like you're cooking the top of your head. To reduce the heat, you can replace the high-wattage infrared heat lamps with lower-wattage flood-light lamps.

Another problem is that the heater bakes the wires in the fixture. Over a long period of time in a corrosive, moisture-laden environment, the insulation on the wires will crack and fall off, which means replacing the whole unit.

But again, if you forego my advice and decide on the fan/light/heater combination unit, choose one that provides significant heat, which can warm a small area quickly. Also try to find a unit that has a glass or a UV-resistant plastic cover over the lights.

INSTALLING A FAN

The method of installing a bathroom fan is pretty much the same for a fan, a fan/light combination unit, and a fan/light/heater combination unit. The only difference is that the combination units have more wires.

Location

The first step is choosing a location for the fan. It should be installed as close as possible to the shower without actually being in the shower (unless it is listed for wet locations and is protected by a ground-fault circuit interrupter). You have to mount the unit between ceiling joists to allow the ductwork to be attached to it, which will affect the placement of the unit. The fixture will come with slide mounting brackets that attach to the ceiling joists (see the photo below).

It's important to cut the hole in the ceiling the correct size because the fan box is designed to sit flush with the ceiling. It has a lip in its frame that fits over the drywall ceiling. In a new installation, you won't have to cut the ceiling, but you still have to install the unit so that the lip will fit over the ceiling material.

Once you've measured, cut the hole in the ceiling, and mounted the fan box, run the cable through the walls (for more on running cable, see Chapter 3). It's best

A typical bathroom fan mounts between joists on adjustable hangers, and the lip on the front edge fits over the drywall.

Manufacturers make a special switch for combination units. For this type of switch, you'll need a box with at least 30 cu. in. of volume.

to bring the power to the switch first and then to the fan. Doing so makes it easier to troubleshoot the circuit.

Switches

There are three basic switching options for bathroom fans: standard switches, dimmers, and timers (for more on wiring switches, see Chapter 3).

If you are installing a fan-only unit, a standard switch will do. If you are installing a combination unit, you can wire each fixture to a different switch, or you can put all of the switches in one box, but you'll need a large-volume box—at least 30 cu. in.—to fit all the wires. Many manufacturers make a special switch for combination units (see the photo at left). If you are putting in a fan/light combination unit, an option is a dual switch—two switches on one yoke.

For combination units, I prefer to wire the light to a dimmer switch. I like to be able to dim the bathroom light because it makes nightly and early morning visits to the bathroom easier on the eyes and can serve as a night-light as well (if one isn't supplied with the unit).

You could also put the fan (or the heater of a combination unit) on timer switches so you don't have to worry about leaving them on.

Wiring

Some units require you to remove the fan module from its box before mounting and wiring (see the left photo below). To do this, loosen the bottom screws and pull the fan assembly out. In one corner of the box you will see a splice box that the fan plugs into—remove the splice box, too (see the right photo below).

To access the splice box, remove the fan module from its metal box. Simply loosen the two bottom screws and slide the fan out.

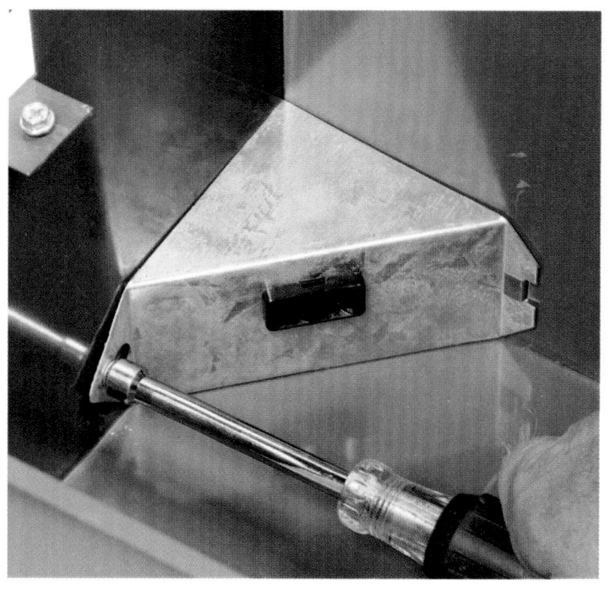

The splice box is normally tucked into one corner of the box. Remove it using a nut driver.

Remove a knockout on the fan box to bring the cable to the splice box. Use a screwdriver to work the knockout loose.

Now remove a knockout on the fan box so that you can bring the cable into the splice box. Insert a screwdriver into the slot on the knockout and wiggle it up and down until it comes out (see the photo above). (The splice box also holds a small, two-prong receptacle that the fan plugs into.) Then insert an NM connector into the knockout (a few NM connectors are shown in the photo at right). Don't just bring the cable in through the sharp bare metal hole because the fan's vibration will cause any sharp edge to cut through the insulation on the cable.

NM connectors protect cables from the sharp edges of knockouts on metal boxes. The threaded end screws through the knockout, the cable is slipped through the connector, and then the screws are tightened to hold the cable in place.

You can now mount the fan by nailing the sliding mounting brackets to the ceiling joists. Make sure the fan box is low enough that the finished ceiling will butt up against the extended lip. In a renova-

Insert the cable through the NM connector and push it into the splice box.

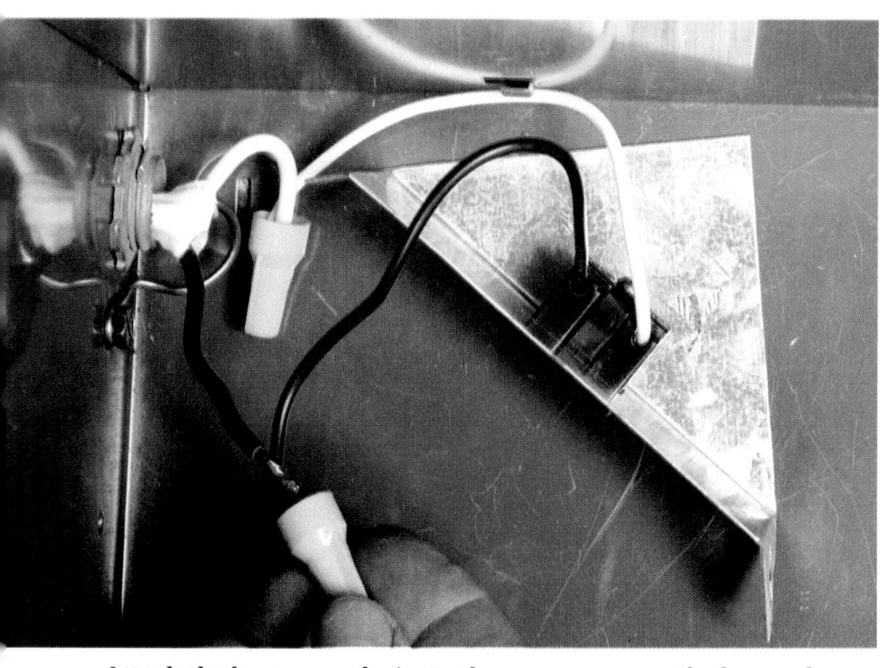

Attach the bare ground wire to the green screw on the box and splice like wires together: black to black and white to white.

tion, this is easy enough. But if the finished ceiling is not up yet, judging this depth could be difficult. For accuracy, slip a small scrap of the finished ceiling, whether it's drywall or wood, under the fan's box to check the fit.

Once the fan box is installed, bring the cable into the splice box through the NM connector (see the photo above). Once in, connect like wires—black to black and white to white—and attach the bare ground wire to the metal box (see the photo at left).

Once the wiring is done, put the splice box back in place, reassemble the fixture, and then test it. If the finished ceiling is not yet up, wait until it is to put the cover

Venting a Bathroom Fan

A common error in bathroom-fan installations is venting the fan straight into the attic. But doing so will send a lot of moisture into the attic spaces and could damage wood members and insulation. You have to vent the fan to the outside, which means running ductwork.

The fan box will have a round (4 in. dia. is common) or rectangular (3¼ in. by 10 in. is common) vent on it, which attaches to the duct. If you have a rectangular vent on the box and want to run round duct, you'll need to install a transition fitting to convert the rectangular to round.

Most fans will come with both side and top vents. If there's an unfinished space above the bathroom, such as an attic, you can use either vent—you can run the duct through the floor space or through the attic. If there's a finished room above the bathroom, you must use the side vent because the duct will have to be run through the floor space.

TYPES OF DUCT

You can use rigid or flexible duct, and both can often be bought as kits. For rigid duct you have the option of using metal or thin-wall PVC drainpipe. Coming in 20-ft. sections, thin-wall drainpipe is smooth on the inside and allows long sweeps for the turns, so the airflow will have minimal disturbance. If metal duct is used, be sure to seal the joints with metal duct tape. In unheated attics, the duct, especially metal, should be insulated to reduce condensation.

The best way to vent the fan—and the easiest to install—is an insulated 6-in.-dia. flexible vinyl pipe that comes in 25-ft. sections. Although there is 4-in.-dia. flexible vinyl pipe available, it should only be used for short runs. I think it's better to use the larger diameter to facilitate airflow (6-in.-dia. pipe has less resistance to airflow than 4-in.-dia. pipe does).

In general, use the 4-in. flexible pipe for runs of 10 ft. or less and 6-in. flexible pipe for runs of 30 ft. or less. For longer runs, use 6-in. rigid duct (metal or PVC).

RUNNING DUCT

For best airflow, try to run the duct in a straight line to the exterior. Don't run it up or down, or make frequent turns. For every significant bend, you will slow down the airflow. By exactly how much I can't tell you, but from my experience, if you have more than four significant turns, you may have to increase the cfm of the fan to provide sufficient air movement. For long, straight runs, try to make the duct slope slightly downward to the outside wall so that any condensed water vapor will drain to the outside.

on. Once you are sure the unit is working, install the ductwork (see the sidebar on p. 101).

Combination units

A fan/light combination unit will mount the same way as the fan unit, but in this instance there will be two small receptacles in the splice box instead of one, and the unit must be wired (if controlled separately by three switches) with three-conductor cable—red, black, white, and bare ground.

Attach the red wire to the light's hot wire (which could be any color, depending on the manufacturer) and the black wire to the fan's hot wire. Also, splice all the white wires together, allowing each fixture to share a common neutral.

A fan/light/heater combination unit (see the photo below) also mounts the same way as the fan unit. With this unit, though, you have to feed at least three fixtures, so you need one hot conductor for each one. Unfortunately, three-

A typical fan/light/heater combination unit has the fan on top, the heater coils to the left, and the light fixture in the center. The light fixture (the white bowl) has a socket for a night-light and another for a standard bulb. The splice box is on the bottom right.

conductor cable has only two hot feeder wires (red and black) and one neutral (white). This means you have to run two three-conductor cables: One for the fan and light and another for the heater (still another may be needed if the unit has a night-light). It would be a good idea to make this a dedicated circuit (see the Glossary on p. 146) because the heater will probably be 120 volts and will draw a lot of current. Never wire a bathroom fan into the bathroom-receptacle circuit—it's not allowed by code.

Sometimes a fan/light/heater combination unit has the splice box located in an easily accessible spot, so you don't have to remove any part of the unit from the box. Also, the splice box may not use receptacles to plug the fixtures in. Instead, each fixture in the unit (fan, light, heater) could be spliced to the incoming power cables within the splice box (see the photo at right). Different manufacturers use different color codes in their wiring schemes, so you'll need to consult the directions or trace out the wires to determine what color wire goes to what fixture.

TROUBLESHOOTING

If you are having problems with a bathroom fan, simply follow the troubleshooting rules for switches and light fixtures (see Chapters 3 and 4).

First check the bulb (if the light is not working), then check the switch and the wiring there. Make sure all splices are tight and verify continuity with a multimeter. If the switch is okay, check the splices within the fan box and check the continuity there. If the problem is not with the wiring, the fan probably needs to be replaced.

If the fan is running but doesn't seem to be moving much air, the problem could be that the seal around the bathroom door is too tight, creating a vacuum within the bathroom. In this instance, try to cut the bottom of the door so there's a 1-in. air gap. This will allow the dry air to be pulled into the room to replace the moisture-laden air being exhausted by the fan.

If you are hearing loud, angry noises coming from the fan (this will sound much different than its typical whirring sound), it means the fan's bearings are getting ready to lock up. The easiest thing to do is to make note of the fan make and model, buy a new one, remove its fan module, and install it in the old unit. This way you don't have to rip out any parts of your ceiling to install a new unit. If your fan dies, and you want to install a whole new fan, try to get one the approximate size of the old one to minimize the carpentry work.

If you have a heater in the unit, and you are having problems, check the wiring around the heater. Make sure it is not melted, burned, or corroded in any area as a result of the heat. If you see corroded, burned wiring, the whole unit will need to be replaced.

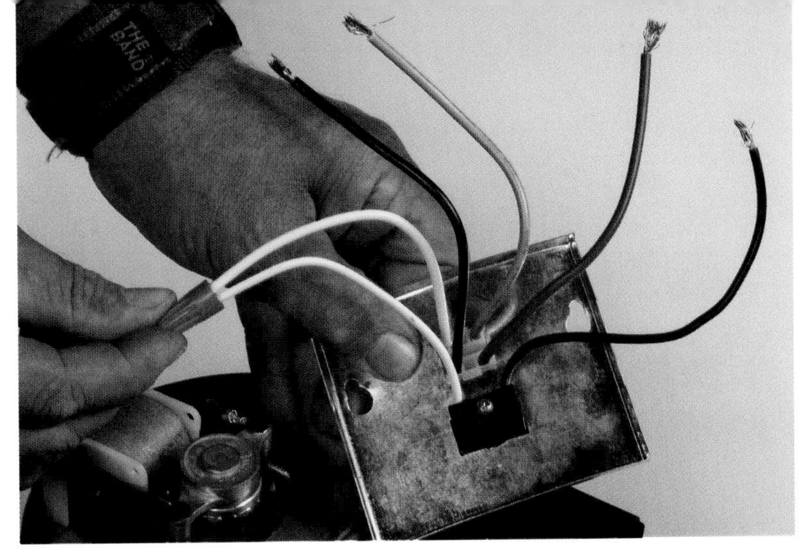

A fan/light/heater combination unit has a lot of wires to splice inside the splice box. What color wire goes to which fixture varies among manufacturers, so you'll need to check the directions to find out which wire goes where.

8

HOME-ENTERTAINMENT SYSTEMS

When I was growing up, people were happy just to have one radio and television in the house. But now home-entertainment systems feature multiple electronic components that need special attention.

Most of the components are tied in together, creating a multipurpose system to delight the eyes and ears. Both TV systems and stereo systems are a lot more complicated than they used to be—even more so than systems that were around 10 years ago.

I go on a lot of service calls regarding home-entertainment systems, and many of the repairs I make could have been done easily by the homeowner. I also have been on many a service call where electronic gear was damaged or destroyed by voltage surges. The gear could have been saved if the installer adequately protected against voltage surges.

In this chapter I'll explain how to choose and run TV cable and how to install a digital satellite system. I'll also explain how to set up and wire a stereo system. And I'll explain how to protect all of this expensive electronic gear against damage from voltage surges. All it takes is a little know-how, a few tools, and a lot of common sense.

TV SYSTEMS

I remember the first television my parents bought. Two delivery people carried it in and set it down on an end table, and I watched in fascination as the first cartoons my eyes had ever seen pranced across the screen. Though I didn't know it then, that box with the flickering light was going to change the world.

Television has come a long way since my childhood. A TV set used to be a luxury. Now it's a fixture in most modern homes. And the technology has evolved as well. Simple black-and-white sets with fuzzy pictures and wavy lines are primitive compared to the advanced televisions of today: Colors are vivid and realistic, the sound quality is clear, and the choice of sizes seems almost limitless.

Television is also becoming more interactive—with both people and other electronic devices in the home—connecting to VCRs, to stereos, and even to the Internet.

Types of cable

One of the biggest advancements in television has been made in cable technology. In the early days of television, the most common TV cable installed in the home was called twin lead—a flat cable about ½ in. wide with one wire on each side covered with plastic insulation. Twin lead is still in use today—if you have an antenna on your rooftop, it probably sends the signal into the home via twin lead.

Twin lead is cheap and transmits the TV signal without much signal loss, but it is not easy to install correctly. Because the wires are only covered with plastic, they can easily be affected by other metal objects that it contacts. Metal can alter the electrical characteristics of the wires and can ultimately impede the signal transmission. Because of this, at the antenna mast, stand-off insulators should be installed every 4 ft. (see the drawing on p. 111) to keep the cable away from the metal mast.

Also, the cable should be twisted twice between each insulator. The twists in the cable minimize signal interference that's coupled into the cable and help stabilize it against wind. To keep lightning off the cable, surge arresters should be installed close to the top of the mast in series with the wires. The arresters will remove lightning-induced voltage surges from the wires and take them to ground.

One of the downsides of using twin lead is that the plastic insulation around the wires cracks easily, exposing the wires inside. It is especially subject to degradation when exposed to the sun's ultraviolet rays. Although modern twin lead is of better quality than the old stuff, and it's available from many electronic stores, the best way to go if you are wiring a TV system is to use coax cable.

Coax Cable

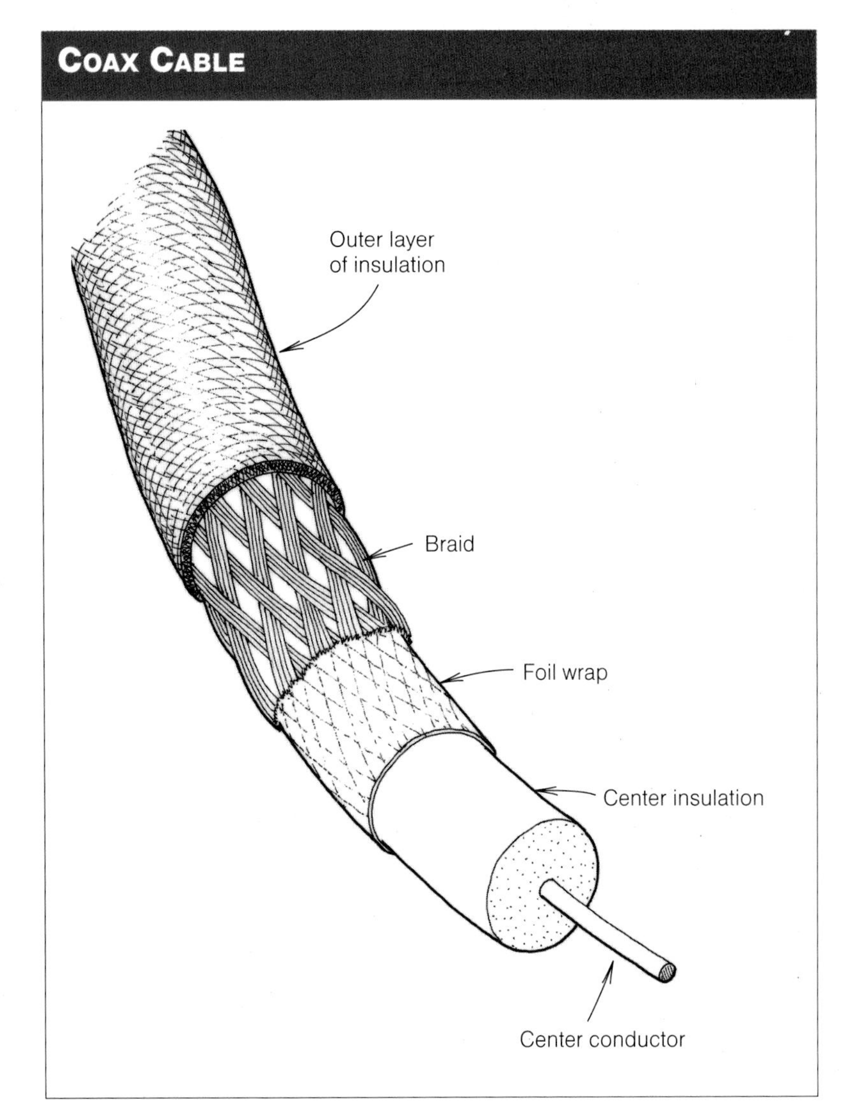

Coax cable has replaced twin lead as the primary TV cable. It is round with one center conductor wrapped in a layer of insulation, which is in turn wrapped with one or two layers of foil, one or two layers of braided aluminum or copper, and an outer layer of insulation (see the drawing above). All of this insulation creates a shield that makes the cable almost impervious to outside interference. You

The female terminals on coax cable are called F-connectors. On the left is a screw-in type, and on the right is a push-in type.

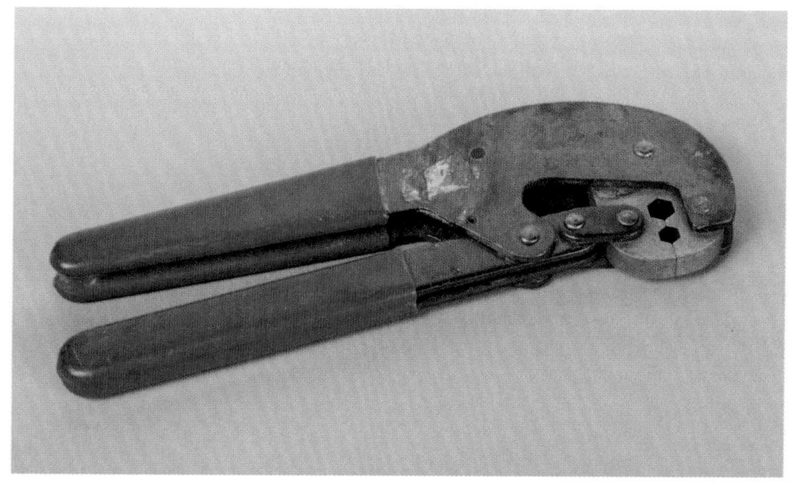

A crimper is used to attach F-connectors onto coax cable.

can run the cable without having to worry about it coming in contact with anything metal.

There are two kinds of coax cable commonly used for video transmission: RG-59U and RG-6. The conductor within RG-59U is smaller in diameter than that in RG-6, so it has a higher impedance (resistance to signal flow). Over long runs, RG-59U could lose signal strength. The conductor within RG-6, on the other hand, is larger in diameter and has a lower impedance, making it less prone to signal loss. RG-6 is the most commonly used coax cable, and it is the cable most manufacturers recommend. (As with electric cable, you cannot run coax cable underground unless the cable is rated for such use.)

The terminals for coax cable are female, called F-connectors. There are two types: One type simply is pushed onto the male and the other is screwed on (see the photo above). As I mentioned, RG-59U and RG-6 have different diameters, so each cable has an F-connector designed specifically for it. Be sure you buy the correct-size connector for the cable you are using.

Coax cable can be purchased in preset lengths with the F-connectors already on them. However, these preset lengths are short, so for long runs, it's best simply to buy the cable by the foot. This way you

Installing an F-Connector

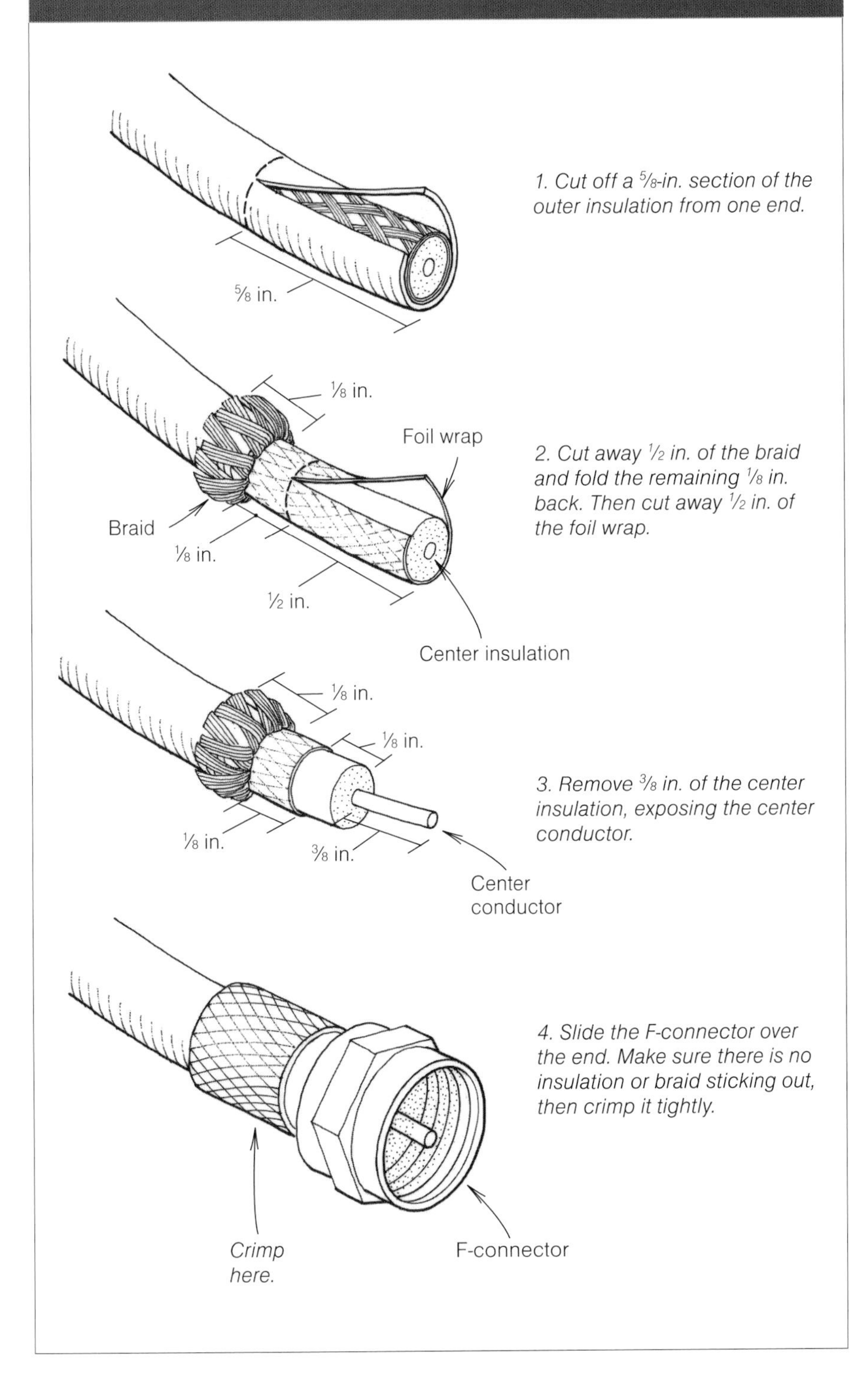

can cut the cable to the length you need and install the F-connectors on it. This is really quite simple to do. All you need is a crimper (see the bottom photo on p. 106), which costs around $20, and the F-connectors, which cost around $2 to $3 for a pack. Both the crimper and the F-connectors are available at most electronics stores.

Although putting an F-connector on a cable end is not difficult, it must be done properly for it to last (see the drawing on p. 107). Once the cable has been run to its location (I'll talk more about running cable in just a bit), use a utility knife to cut away about ⅝ in. from the end of the outer layer of insulation. Cut away about ½ in. of the braid and fold back the

Distributing TV Cable

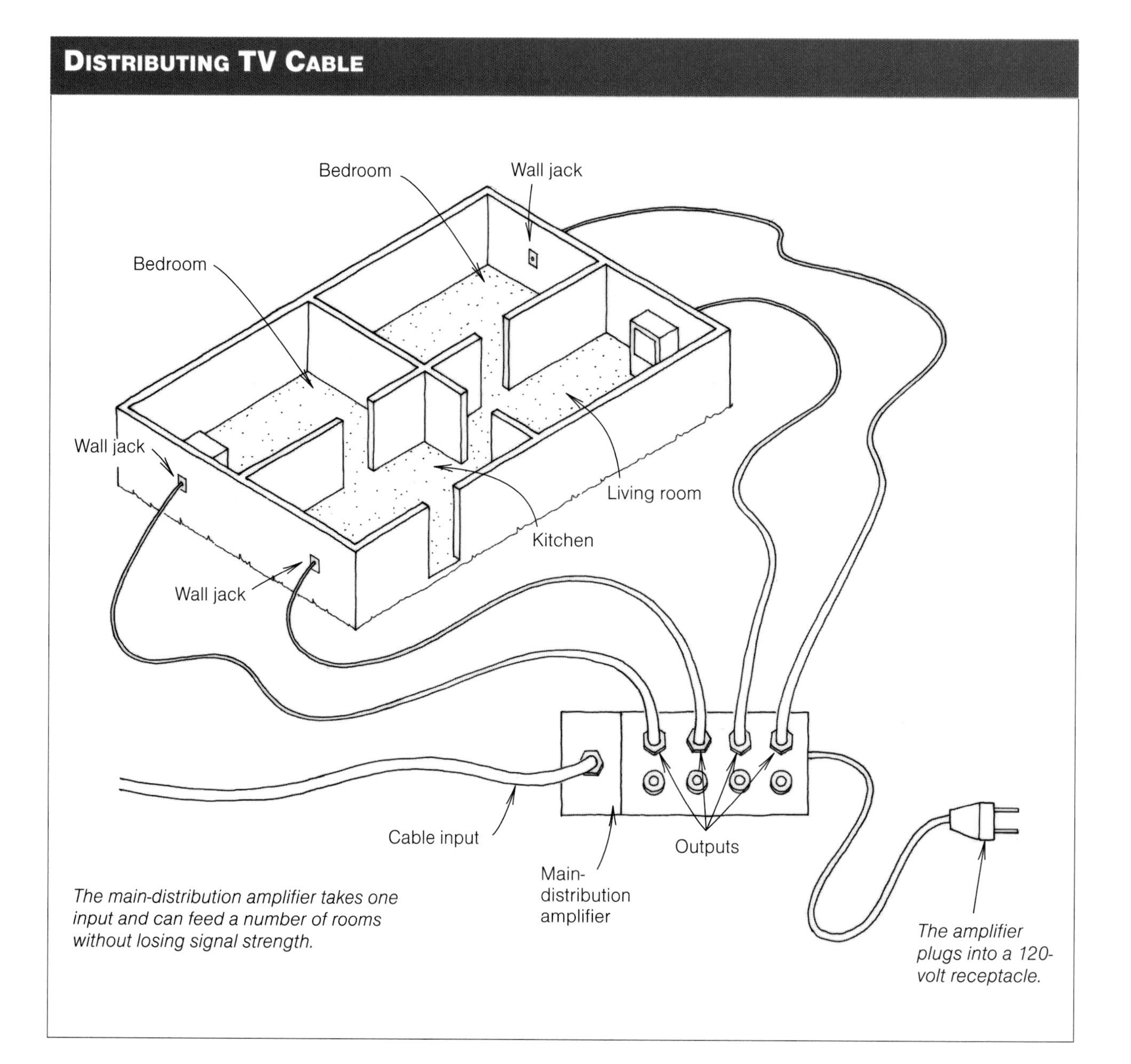

The main-distribution amplifier takes one input and can feed a number of rooms without losing signal strength.

The amplifier plugs into a 120-volt receptacle.

remaining ⅛ in. Then remove ½ in. of the foil wrap and ⅜ in. of the center insulation.

The copper conductor in the center should be exposed for ⅜ in. Now slide the F-connector over the end, making sure there is no insulation or braid sticking out. Crimp it tightly on the cable, and you are ready to connect.

Running cable Although many people don't know it and often spend money having a professional wire their homes for cable, it is legal and very common to wire your own home. Whether your house is going to receive broadcasts from a local cable company, from an antenna on the rooftop, or from any type of satellite dish, the interior cable runs will all be the same. (This discussion will assume the use of coax cable.)

Coax cable can be run easily along or through studs and joists because it is flexible—not stiff like 12-gauge electrical cable—so it makes turns with ease. Simply pull the cable to where it needs to go. However, make sure the cable does not get kinked because a sharp bend could interrupt the signal. Also, don't run it near heat sources, such as flue pipes, because the heat could damage the insulation.

Unlike electrical cables, there are no official requirements about stapling coax cable, but there are a few commonsense guidelines. When running the cable along studs or joists, support it with staples every 2 ft. or 3 ft. (staples for coax cable can be purchased at electronics stores). But don't drive the staple in too far, or it could cut through the cable's insulation and damage the center conductor.

A splitter has one input and two outputs to feed two televisions.

If only one television is to be connected, the cable can go straight to that location. But if there are several televisions or if you'd like to install a jack (a cable outlet) in every room for future use, it's best to wire all rooms from one central location. To ensure that the signal will stay strong enough to be used at all locations, a well-designed system will start with a main-distribution amplifier where the coax cable enters the house (see the drawing on the facing page). The amplifier increases the signal strength and is plugged into a standard 120-volt receptacle. This type of amplifier has several outputs on it (it could have as many as 12) with one input.

If you want to feed only two locations, or jacks, you can install a splitter anywhere along the cable's run. A splitter is a connecting device that has one input and two outputs (see the photo above). Simply attach the feeder cable to the input terminal and the output cables to the other terminals. The bad thing about using a splitter, however, is that you lose

Grounding Coax Cable

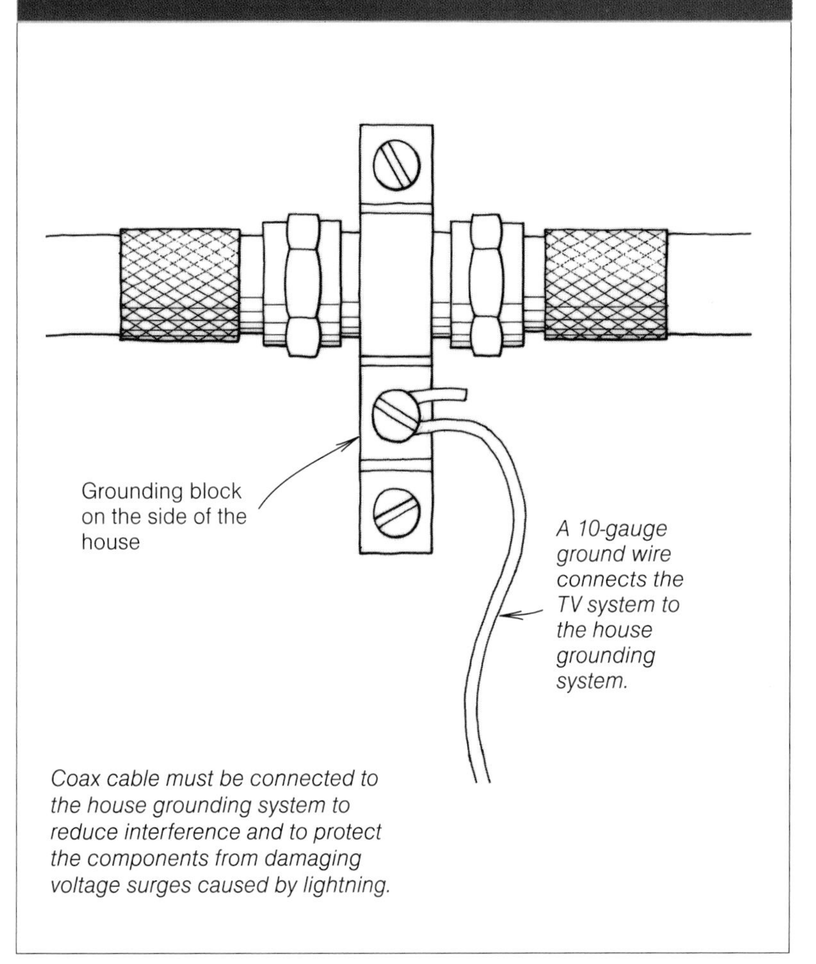

Coax cable must be connected to the house grounding system to reduce interference and to protect the components from damaging voltage surges caused by lightning.

a little signal strength over long runs. To retain the signal strength, use a splitter that's equipped with an amplifier, which, of course, is more expensive than a standard splitter.

Run each cable to a wall jack, which should have its own rough-in box and be protected by a cover plate. The rough-in box is typically a standard small-volume electrical box (18 cu. in.). It is okay to share a rough-in box with a phone jack, but never install a cable jack in the same box with electrical wiring, even if your local codes approve. Some codes say it is okay as long as an approved separator (a plastic wall) keeps them apart. But one accidental touch of 120 volts to the low-voltage TV cable will blow out everything on the system.

All coax cable must be connected to the house grounding system to reduce interference and to protect the components from damaging voltage surges caused by lightning. To ground the cable, attach a grounding block on the outside of the building where the cable first comes to the house (see the drawing at left). From the grounding block, run a 10-gauge bare copper ground wire to the grounding electrode conductor and attach it with a split-bolt connector (available at electrical supply stores). Do not attach the ground wire from the grounding block to a ground rod that is not connected to the house grounding system because it is a code violation. The NEC says a house can have only one grounding system.

Antennas

Along with advancements in how televisions function and look, and in the cables that are used to connect them, even greater advancements have been made in how the television receives its signal.

In the early days of television and even today, giant transmitting towers with glowing lights broadcast TV signals to be received by antennas mounted on rooftops. These old-style TV antennas, called Yagis (see the drawing on the facing page), look like flattened porcupines and are still in use in many areas of the country where cable is unavailable (or for folks who simply don't want to pay for television).

Grounding a TV Antenna

An ungrounded antenna and the metal mast it mounts on can build up high voltage surges during a lightning storm. I've seen it happen. Before I had the opportunity to ground the antenna on my house, my area was hit with a tremendous lightning storm.

During the storm, I heard a loud ZAP and saw a blinding flash from behind my stereo gear. I knew immediately that my equipment was destroyed. I could have prevented this damage—and saved some money—by grounding the antenna and mast.

To ground an antenna, run a 10-gauge bare copper wire from the metal mast directly into the house grounding system. Attach the wire to the mast with a large pipe clamp approved for outdoor use (available at electronics stores and electrical supply stores). Then connect the wire to the grounding electrode conductor using a split-bolt connector (available at electrical supply stores).

Do not attempt to ground the antenna by connecting the wire to a metal water pipe. This will simply put the voltage surge on the pipes. Also, although you can buy tiny ground rods at electronics stores that can be attached to an antenna, don't use them, no matter how tempting. A ground rod that is separate from the house grounding system is a code violation—the house is allowed only one grounding system—and it's dangerous.

If the ground wire is not connected to the house grounding system, and a hot wire touches the mast or antenna, the breaker will not kick off. Anyone touching the mast or antenna in this situation will be electrocuted.

Old-Style Antenna

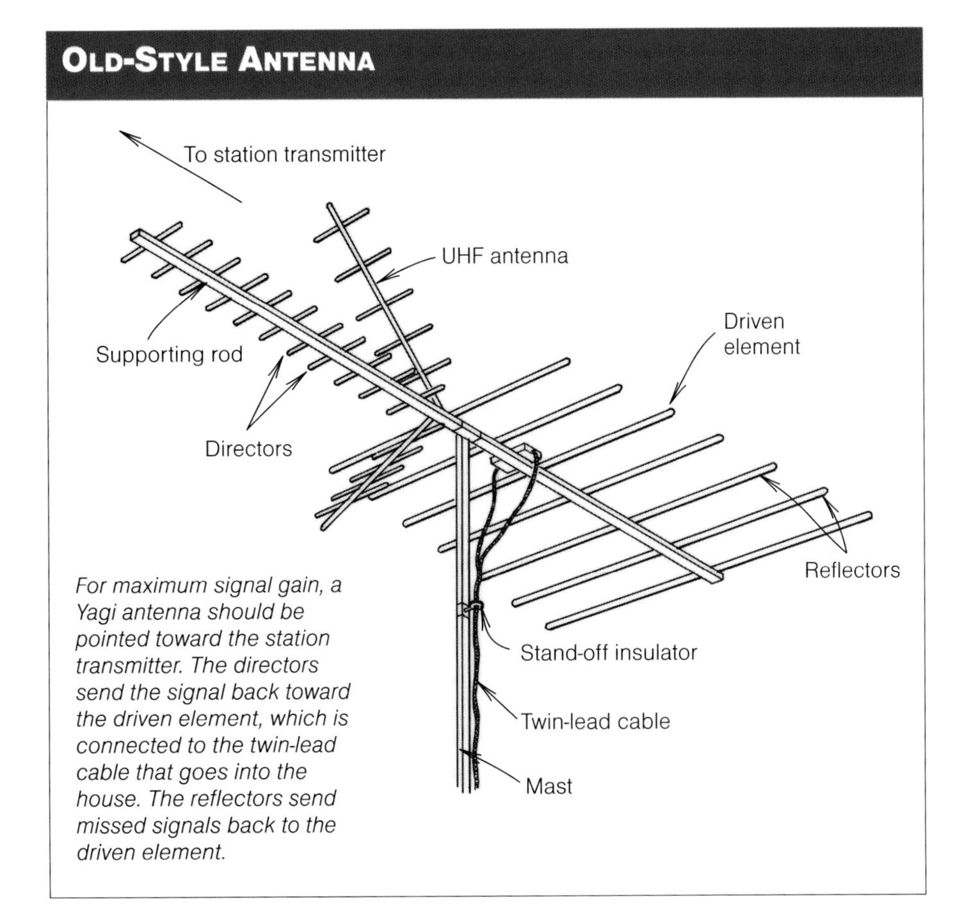

For maximum signal gain, a Yagi antenna should be pointed toward the station transmitter. The directors send the signal back toward the driven element, which is connected to the twin-lead cable that goes into the house. The reflectors send missed signals back to the driven element.

Using an antenna allows you to get free reception, but the quality is sometimes poor, and the coverage area that the signals can reach is limited, which means that many of the signals could be weak. If the signal is weak, you'll be watching more snow (white dots all over the screen) than programs.

The transmission signal is comprised of very-high-frequency (VHF) and ultra-high-frequency (UHF) channels. Some antennas can receive only one or the other, but some can receive both. The long, flat part of an antenna that's parallel to the earth picks up the VHF signal. The V-shaped part of an antenna picks up the UHF signal.

The signal-gathering ability of an antenna, called signal gain, makes the difference between seeing a good, crisp picture and seeing one littered with snow. In general, the larger the TV antenna, the more signal gain it has.

A TV antenna is composed of a large supporting rod down the center that holds several smaller, round metal rods called directors, which direct the signal toward the driven element. The driven element collects the signal, then sends it through the lead-in transmission line (typically twin-lead cable), which sends it to the television. The more driven elements an antenna has, the better the signal gain. Behind the driven element are reflectors, which reflect any missed signals back into the driven element to boost the signal strength.

For optimum reception, install the antenna as high as possible, but keep it away from power lines, which can interfere with the signal.

Though an antenna has the ability to receive several channels or stations at once, most of the time the transmitting towers will be too far apart for the antenna to effectively receive all of them. For the best-possible signal gain, the thinner end of the antenna should be pointed toward the transmitter of the TV station you want to pick up. This can either be done manually by climbing onto the roof and twisting the antenna mast or automatically with a motorized antenna rotator (commonly available for around $100). The antenna controller is located at the TV set, and the antenna is mounted on top of the motor. Simply turn the controller's dial toward the direction you want the antenna to turn to, and the antenna will start moving. You'll know the antenna is pointing in the right direction when the picture on the television is best.

If you find that your signal is weak no matter which direction you turn the antenna, you can install a signal booster, which is an amplifier that increases the size of the received signal. Since the amplifier will increase electrical noise (all electrical circuits have noise) as well as the signal gain, it must be bolted to the TV antenna mast far from the television. If the amplifier is installed closer to the television, it could pick up and amplify noise from phone lines and electrical appliances that will transmit through the television.

Satellite systems

Antennas are primitive compared to today's methods of sending and receiving TV signals. Most homes receive TV signals from cable TV companies through cables mounted on poles outside the home. But satellite technology is quickly gaining momentum because of its fantastic picture quality.

This satellite dish was considered state of the art just a few years ago. But it's now obsolete due to the introduction of smaller, more-powerful digital satellite systems that can be mounted onto the house.

When I bought a 10-ft. satellite dish and installed it in my yard a few years ago (see the photo above), it was considered state of the art, although it was expensive at the time (a couple thousand dollars). The dish receives C-band signals from around 20 satellites. Because C-band signals have a low frequency and limited transmitting power, a large dish is needed. But one of the biggest satellite dealers in my area of Virginia, which used to install over 100 of these large dishes a month, is now installing zero. The owner says this technology is dead.

Digital satellite systems (DSS) are now the current technological rage and offer two distinct advantages over C-band systems. First, DSS dishes are small, 18 in. to 39 in. in diameter, and are less expensive than the dishes needed to receive C-band signals. Plus DSS dishes can be mounted directly onto the house.

Second, the DSS system receives high-powered Ku-band signals in digital form directly from satellites. The Ku band is a much stronger signal than a C-band signal, with higher frequencies. And because it receives the signals in digital form, DSS provides a crisp, clear picture and full-spectrum sound. Most dealers have a system up and running in their showroom to impress you as you walk in (to find a dealer near you, check the yellow pages under "Satellites"). And it is impressive—no snow, just pure, deep colors.

The cost of the system will vary, depending on the package you buy. You can probably get a complete system, including the dish and the receiver (which decodes the digital signal), for around $400. But that money is just for the equipment.

You also will pay a monthly programming fee, and that figure will depend on the package of channels you buy. Although most services offer many of the popular channels, before you buy and install a DSS system, check with the dealer to be sure you'll be able to watch your favorite programs. It's also a good idea to inquire about any additional fees for first-run movies or special events. If you want to watch local programming, you'll need to hook up your rooftop antenna to the system.

Although 99% of the time, the picture will be crystal clear, Mother Nature occasionally interferes with the signal. If your area receives very heavy rain, the signal gain could be reduced. It's possible to lose as much as 20 minutes of signal during a week of heavy rain.

Installing a DSS system The DSS dish must be mounted outside to pick up signals from the satellites. In theory, the dish can be easily installed by a handy do-it-yourselfer. However, the installation manual is around 1 in. thick, with a lot of small details. If you desire, most dealers offer service contracts that will include installation and maintenance of the system. But this will cost extra. If you prefer to install the dish yourself, follow all manufacturer instructions and warnings. Here are the basics.

Securely mount the dish to a solid surface, preferably into the house framing. Be sure the mast is plumb—exactly straight up and down—or the system will not work properly. Check for plumb using a level or a plumb bob. Because the satellites that transmit the signals orbit the equator, you have to face the dish toward south. Once the dish has been mounted, make all the cable connections.

Run a single RG-6 coax cable from the DSS dish to the grounding block mounted on the outside of the house. Before attaching the cable to the grounding block, run a 10-gauge bare copper wire from the block to the house grounding system. If the cable from the dish or the cable to the house is coming to the grounding block from above, form it into a drip loop to prevent water from running down the cable and seeping into the connections (see the drawing on the facing page). Simply make a small loop in the cable just before the grounding block and secure it with a tie wrap. If the cable goes straight down as it leaves the grounding block, a drip loop isn't needed.

Making Drip Loops

If the cable connects to the grounding block from above, drip loops should be formed at the grounding block. The drip loops prevent rain and condensation from getting into the cable connections.

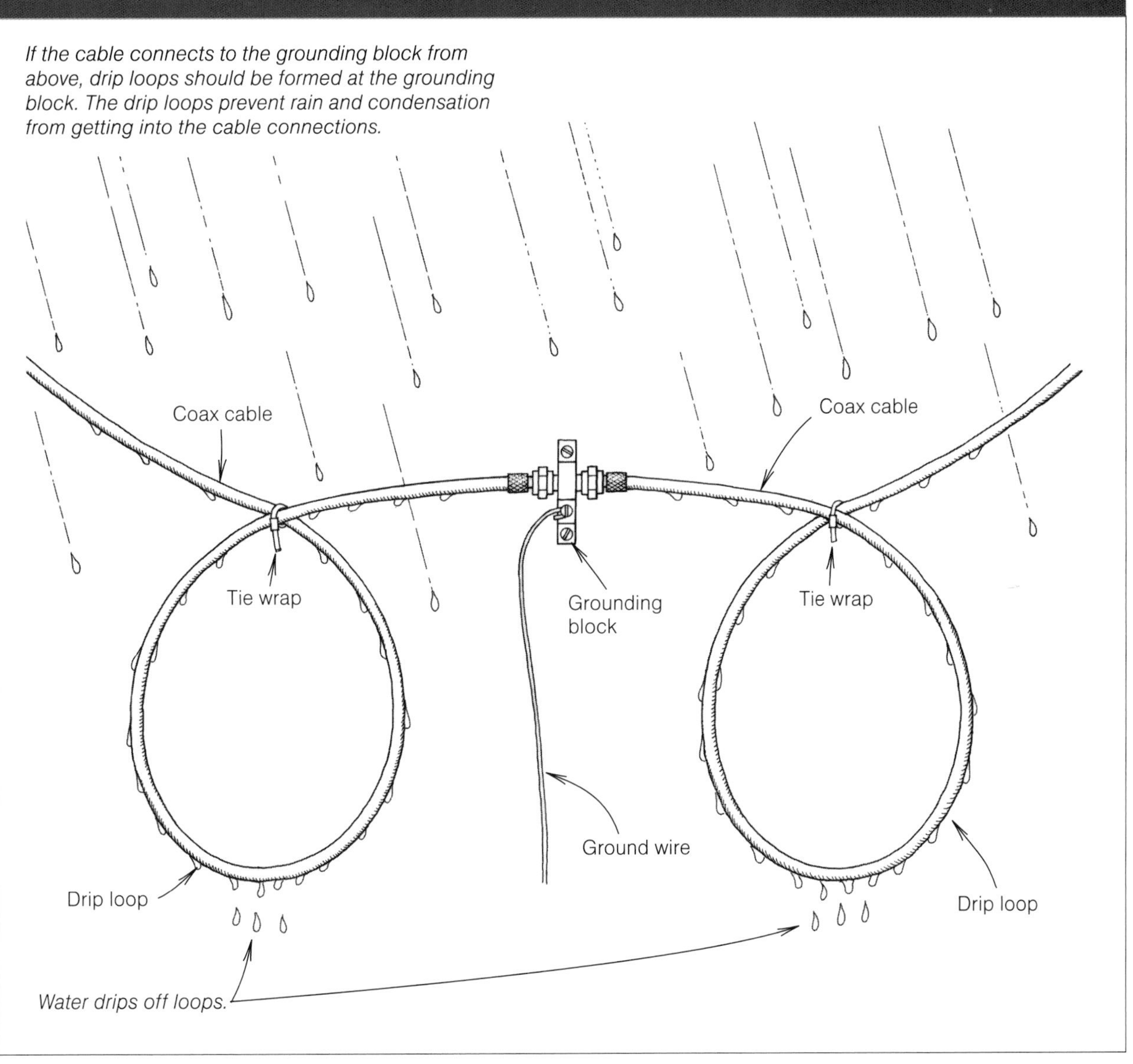

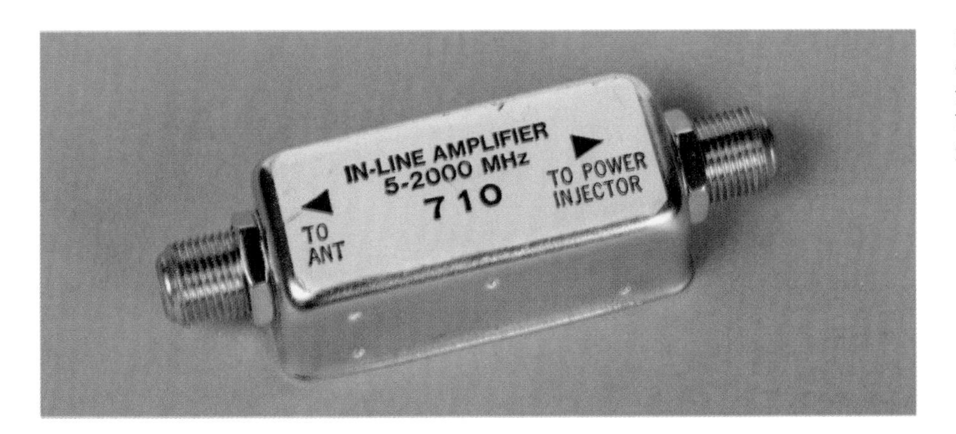

If the distance from the DSS dish to the receiver is more than 100 ft., you'll need to boost the signal with an in-line amp like this one.

Typical DSS System

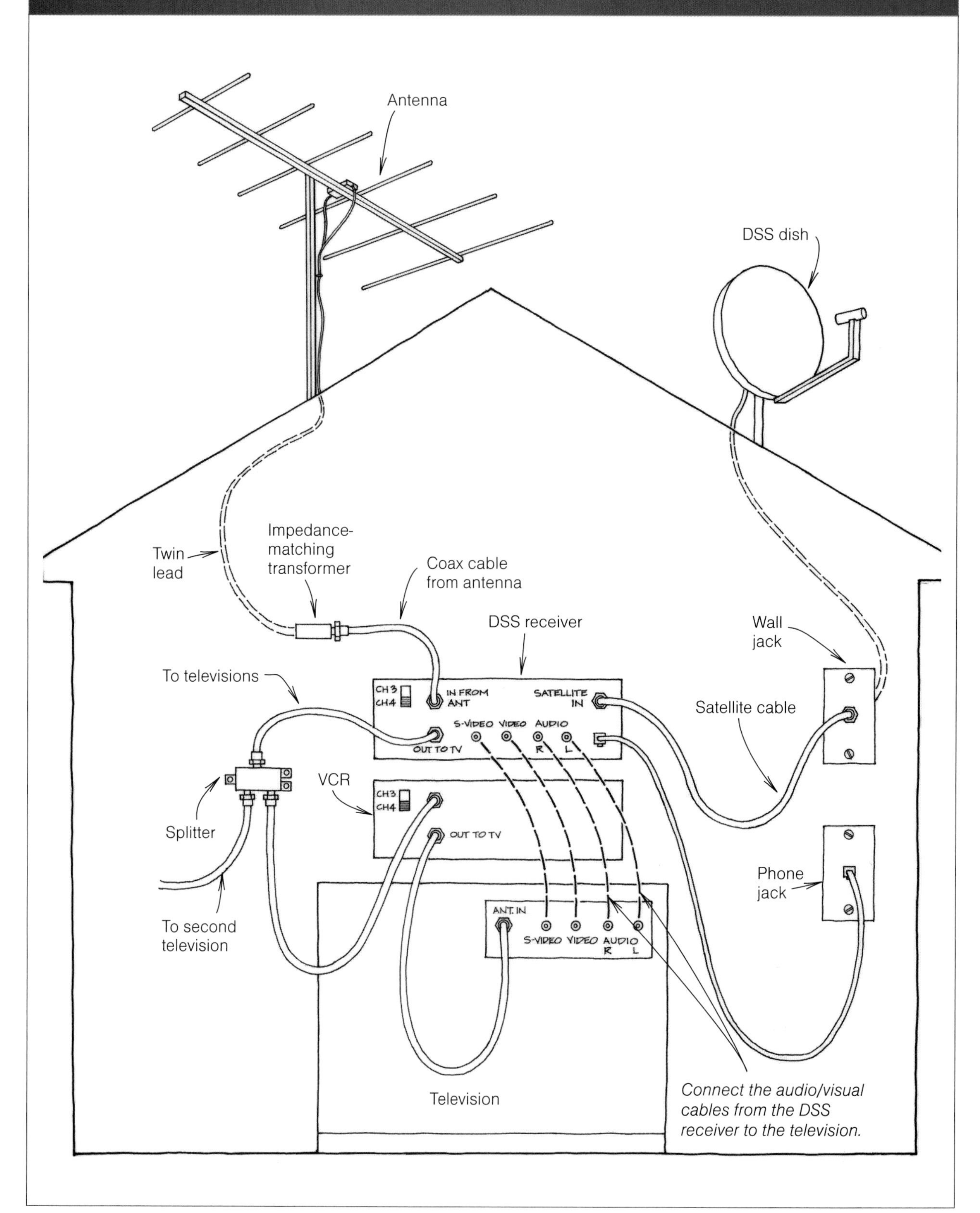

Connect the audio/visual cables from the DSS receiver to the television.

The length of the cable run from the DSS dish to the receiver is critical. If the run from the dish to the receiver is excessive (more than 100 ft.), the signal could be drastically reduced. In this instance, you'll need to install an in-line amplifier at the dish to boost the signal (see the photo on p. 115).

From the grounding block, run the coax cable into the house and to a jack in the TV room (see the drawing on the facing page). It doesn't really matter where you bring the cable into the house. Just drill the correct-size hole (about ⅜ in.) in the wall and pull the cable through. Once you are through, seal around the hole with caulk to prevent insects, moisture, and drafts from entering the house.

Bring the cable to a wall jack and connect the "satellite in" jack on the receiver to the wall jack. Then connect the receiver to a phone jack (billing is done via the phone lines).

The connection from the dish to the receiver must be direct. You cannot install a splitter before the receiver because it must first decode the signal. If you want to feed two televisions, the splitter must be installed between the receiver and the TV sets. If you want to feed more than one television and want to watch two different channels at the same time (one person in the bedroom watches one channel while another in the living room watches another), you will need two of everything: two LNBs (the thing that catches the signal at the dish), two receivers, as well as two RG-6 cables (one for each receiver connection). The option is nice, but the expense is doubled.

Some people like to watch local programming, but DSS systems usually don't carry it. But this is easily fixed by connecting the DSS receiver to the Yagi antenna on the roof. If your TV-antenna cable is twin lead, you'll need to connect it to coax cable to connect to the receiver.

The problem is that the two cables have different impedances (see the Glossary on p. 146): coax has an impedance of 75 ohms while twin lead normally has an impedance of 300 ohms. To make the splice, you'll need a special device called an impedance-matching transformer, which is sold at all TV and electronics stores. An impedance-matching transformer balances the incoming and outgoing impedances of the different cables to maintain the signal strength.

Aligning the dish Once all the attachments have been made, the dish has to be aligned correctly. Proper alignment is critical to the performance of the system, and it is impossible to align the dish by just pointing it toward the general direction of the satellite orbiting the equator.

The dish has two alignment adjustments: elevation and azimuth. Elevation is the height and angle adjustment of the dish. Azimuth is the left and right adjustment—technically the amount of degrees in the clockwise direction from true north. The manual that comes with the system provides complicated instructions for finding the correct elevation and azimuth. But with most systems, you can simply turn on the receiver and call up a screen menu that automatically gives you these numbers (consult the manual for specific instructions). Enter your zip code, and bingo, the figures appear on screen.

Creating a Home Theater

Home theater puts the movie action right into your living room by giving you total surround sound. You are not just a spectator—you are part of the experience. War planes enter through your kitchen, blaze a path across your head, and crash right in front of you. Monsters jump out of the screen and sit in your lap.

Home theater can be created on almost any budget. If you already have a stereo system, you need to add three more small speakers and an audio/video decoder amplifier with Dolby sound. The amplifier, which will run about $300 and up, decodes the Dolby signals and powers the extra speakers.

The amplifier has four channels (there are now systems available that offer five independent channels, but they are expensive): front left, center, front right, and rear. For the best sound, place your regular speakers in front of the viewing area, on either side of the television (see the drawing below).

Locate one speaker directly in front of the viewing area on top of the television. Because a regular speaker generates a magnetic field that will adversely affect the television signal, you can't put it on top of the TV set. You must use a speaker that is magnetically shielded. You'll find magnetically shielded speakers specifically designed for home theater at electronics stores.

Place two other speakers behind the viewing area for depth. If you want another level of sound, add a subwoofer. The subwoofer can be located anywhere in the room, but for best effect, center it directly behind the viewing area (if the area is open).

The center speaker in front carries the sound from left to right or vice versa. The small speakers in the rear add depth, and the subwoofer accents the whole sound with additional base.

With this setup, the illusion of movement is created because the sound follows the action on the screen. For example, if the screen shows a car driving off to the right, the sound will move with the car, starting at the center speaker and moving to the right speaker. When a plane roars across the screen from right to left and then crashes, you will hear it enter from the right speaker, go across the center speaker, and crash on the left speaker. Even monsters will reach out and grab you.

Home-theater layout

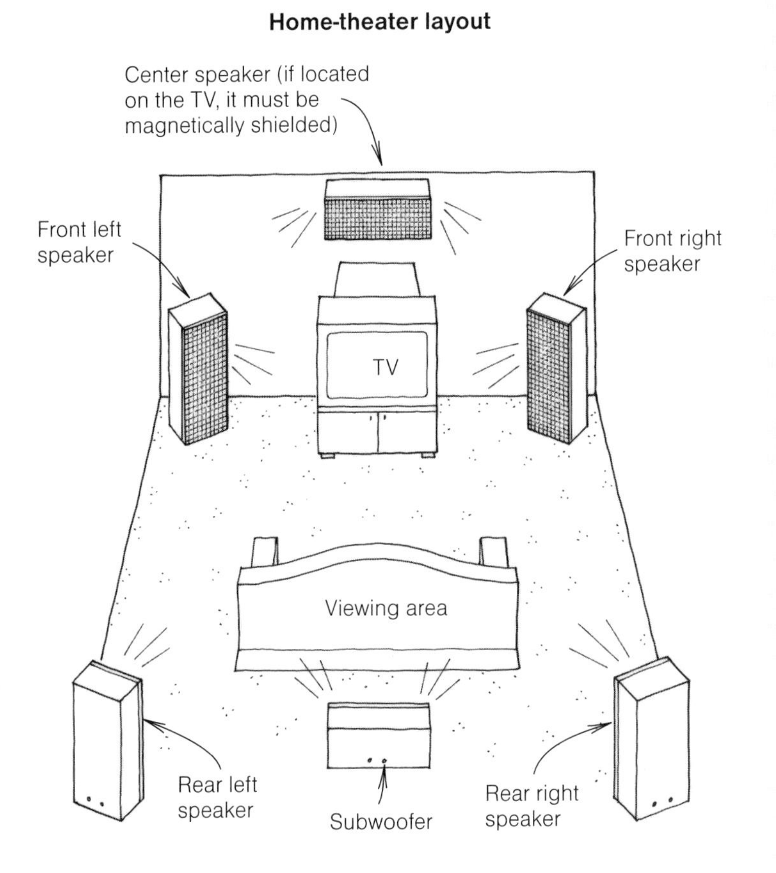

To set the elevation, loosen the elevation nut on the dish (see the instructions for the location of this nut) and move the dish up or down to the elevation setting given on the screen. The typical elevation setting for most systems is 32°, but check the manual.

To align the dish to the azimuth setting (in degrees), you'll need the help of a compass (which is typically provided with the installation kit). Loosen any nuts holding the dish arm tight and then use the compass to show you the direction of the required azimuth setting. Point the dish in that direction, then tighten the nuts. You should have a picture now, but it might not be great.

The elevation and azimuth settings given by the manufacturer are coarse adjustments only—they simply allow you to get the signal from the satellite. To get the maximum-strength signal, you'll have to fine-tune the dish's alignment. Consult with your manual on how to fine-tune the unit.

Troubleshooting

If you have poor TV reception, there are a few troubleshooting steps you can take before making an appointment for what could be an expensive service call.

If you have a DSS system, it will basically troubleshoot itself. Simply follow the directions in the manual. If you have an antenna as the main signal receiver, and your picture quality is not good, check to be sure that the antenna is pointed in the right direction. This, unfortunately, will require a trip to the rooftop. Most often, however, poor reception can be traced to the cable.

If you have an antenna, make sure that the twin-lead cable is in good condition. The older styles don't last very long. If you see the cable is damaged, replace it.

Replacing a Damaged Section of Coax Cable

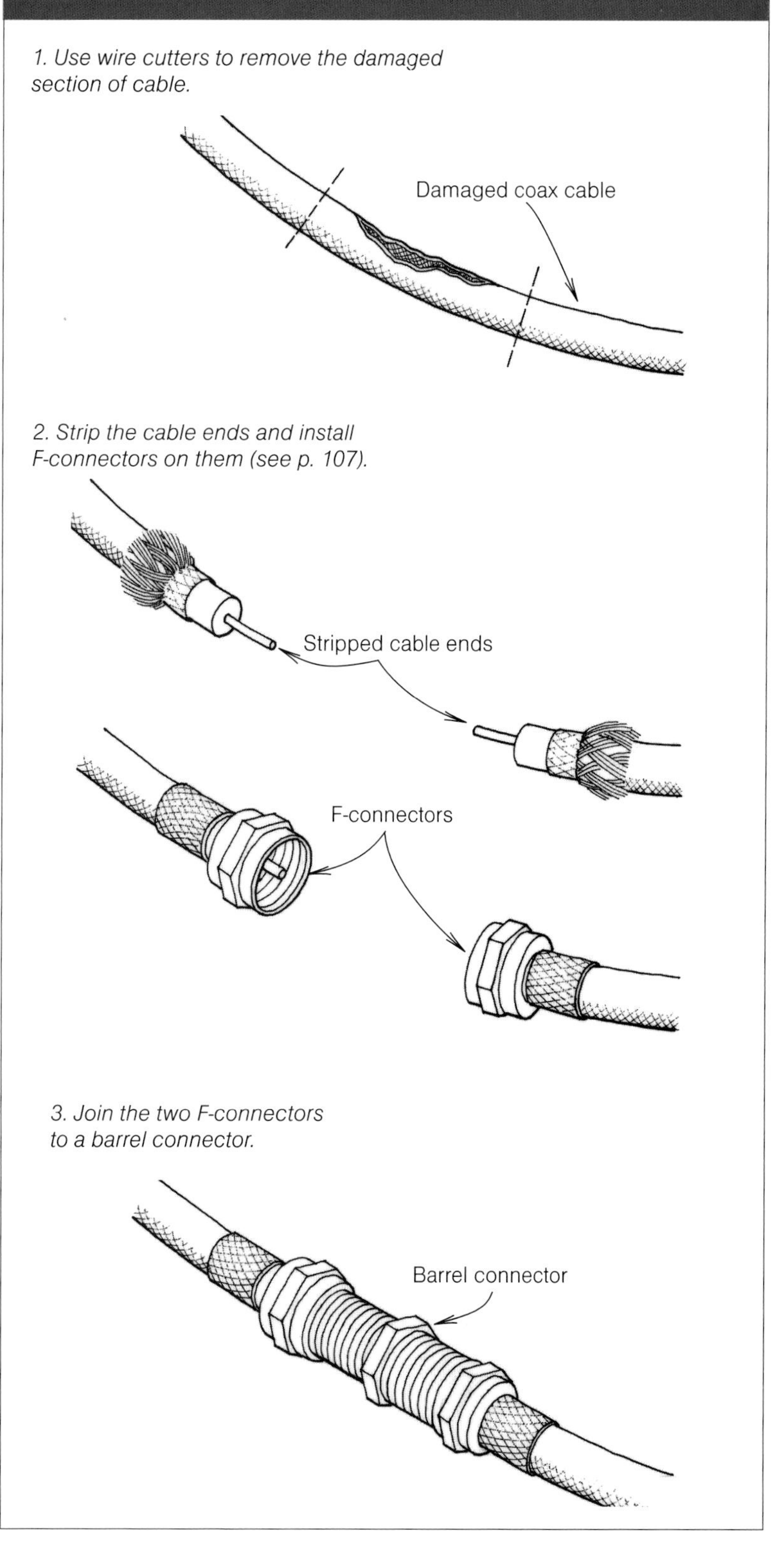

Also check to be sure that twin lead is not directly connected to coax cable at any point. This splice requires the use of an impedance-matching transformer to balance the impedance of the cables. I had a customer who had complained of a poor picture since she moved into her new house. I found that the antenna installer had run twin lead from the antenna to the coax jacks in the wall. The impedance mismatch caused the signal strength to degrade, resulting in a poor TV picture.

There are fewer problems with modern coax cable than there are with twin lead. If your are experiencing trouble with your TV reception, check the cable runs. Make sure the cable is not kinked or run near heated objects such as flue pipes that extend through the attic. The heat could damage the cable. Also, if the cable run is long, resulting in a snowy picture, install an in-line amplifier.

Look for damaged cable and loose connections. Installers sometimes are working so fast that they drive a staple through the cable instead of around it. Other times the staples are driven too tightly, crushing the cable. If any F-connectors are loose, replace them.

If you find that a section of coax cable is damaged, you don't have to replace the whole thing. You can remove the damaged area and make a splice with a barrel connector (available at electronics stores) between two F-connectors (see the drawing on p. 119).

If none of your repairs have solved the problem, you'll need to call the cable TV supplier or your satellite dealer to arrange for a service call.

STEREO SYSTEMS

Because of the number of elements that make up a modern stereo system, and because the stereo system is also tied into other electronic components, such as televisions and VCRs, care must be taken in the organization and wiring of components.

If you have a substantial stereo system, it's a good idea to wire a particular room to handle the load. Bring in extra circuits for power, and arrange the components so that you can make the connections easily.

Also, because electronic components generate heat and will not work properly if they are too hot, be sure that you provide ventilation. If the gear is within a cabinet with a door, cut 2-in. vent holes in the back of the cabinet to allow air circulation. Cut four holes in the cabinet's back—two at the top and two at the bottom—more if you need to. If you stack components, provide plenty of airspace between levels.

Wiring speakers

Wiring stereo speakers is really pretty simple. If you are attaching two speakers, you'll need two cables—one cable for each speaker—each with two 14- or 16-gauge wires.

The stereo or receiver output will probably have two channels: A and B. Each channel, labeled left and right, can be wired to a set of speakers. These can be heard as A only, as B only, and as A and B together. Run two cables from a channel and connect one cable to each speaker. For extra base, you can add subwoofers to the channel that's not used.

Wiring Stereo Speakers

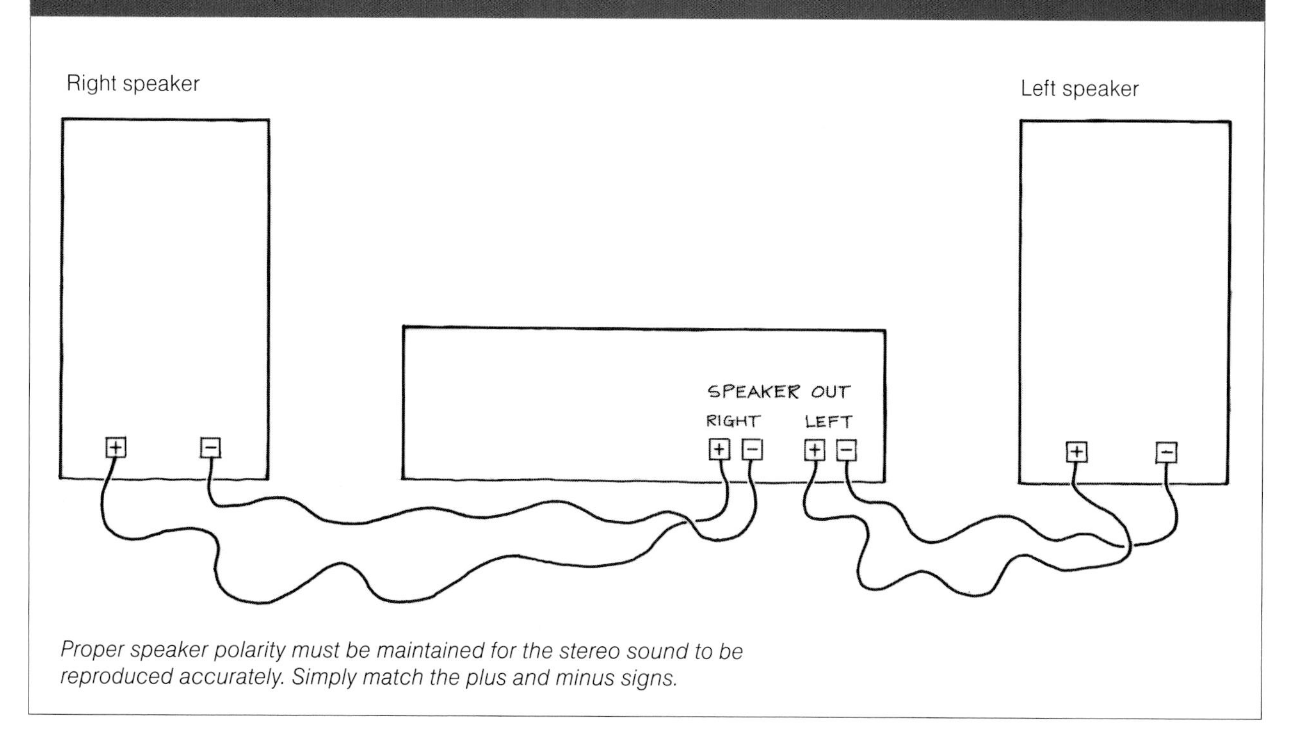

Proper speaker polarity must be maintained for the stereo sound to be reproduced accurately. Simply match the plus and minus signs.

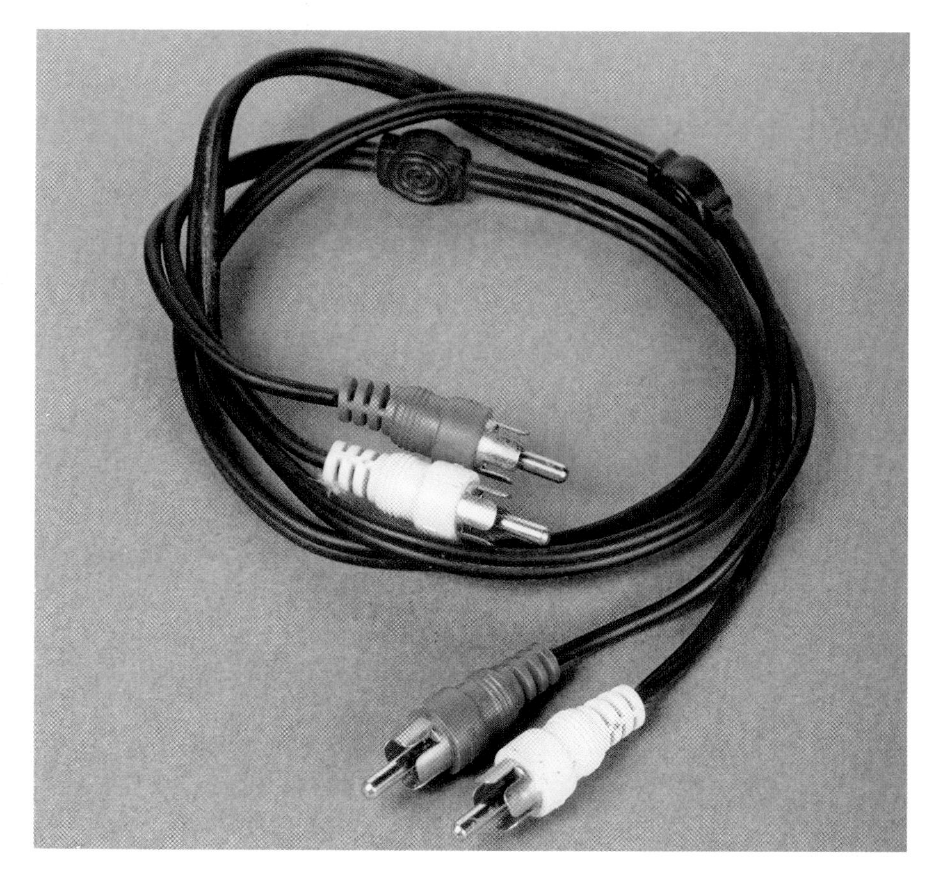

RCA interconnect cables are the most common cables used to connect stereo equipment. Use a different color for each component so that you don't mix up the cables at the receiver or amplifier.

For speakers to work properly, they must be wired with the proper polarity. That means the positive output (+) on the receiver must be connected to the positive input on the speaker, and the negative output (–) on the receiver must be connected to the negative input on the speaker (see the drawing on p. 121). Most speaker cables are color coded or ribbed to help wiring to the correct polarity.

Connecting components

Virtually all stereo components are connected via RCA interconnect cables (see the photo on p. 121). If you have a lot of connections to make, and you'd rather not have a dozen single cables lying around looking like a bird's nest, you can get this type of cable in double and quad versions, which provide two and four hookups, respectively.

Long cables could reduce the signal gain, so try to keep all the components as close as possible but still allowing for airflow to dissipate heat. Connections are simple because components typically have the jacks labeled (for CD players, tape players, etc.). Just be sure that you don't run the cables in traffic areas or near heat sources.

When hooking up the components, it's a good idea to use different color cables for each component or tape or label each cable. This way you don't mix up the cables at the receiver or amplifier. Doing so will not allow the system to function properly or at all.

If you want to hook up your VCR to the stereo system for better sound quality from your television, use RCA interconnect cables. On the VCR, insert the cable into the audio out jacks (there will be a left and right channel). Run that cable to the receiver or an audio/video amplifier. On a receiver, connect the cable to the auxiliary jacks (connect the left and right outputs to their corresponding jacks on the input). On an audio/video amplifier, connect the cable to the VCR input.

Troubleshooting

When it comes to stereo systems, there's not much to troubleshoot. All electronic components come with complete installation and troubleshooting instructions. If you are experiencing troubles with your system, consult your manual for specific directions. If a component is not working at all, you may need to send it out to be repaired.

In general, though, the biggest mistake people make when hooking up components is mixing up the RCA and speaker cables. Doing so will not allow the system to perform well, and the result will more than likely be poor sound reproduction. If the sound is distorted, double-check to be sure that all the connections are properly polarized.

SURGE PROTECTION

Voltage surges are a major source of damage to today's sensitive and expensive electronic components and appliances. The two most common ways that voltages are induced into the wiring are through lightning (see the sidebar on the facing page) and through other appliances.

One customer saw a ball of lightning shoot out of a receptacle behind his stereo, flash across the room, and fly into the metal flue of his woodstove. The lightning destroyed his receiver but,

How Lightning Gets into the House

The majority of lightning-induced voltage surges come into a house via the utility and telephone lines and seemingly head for the most expensive and sensitive electronic equipment, such as computers, stereo systems, and televisions.

Lightning produces magnetic lines of force that couple through the air, inducing voltages on both current-carrying conductors (such as utility lines or in-house wiring) and non-current-carrying conductors (such as metal plumbing pipes and appliance frames). The lighting does not have to hit or even come close to the house to have an effect on it.

The lightning can hit many miles away, putting a damaging voltage spike into the power lines that travels through the service drop right into the home (see the right drawing below). Or a lightning bolt could flash across the sky above the house and induce voltages within the house into anything that will conduct electricity (see the left drawing below).

The ground connections on the utility poles often are only good enough to lower voltage surges, not stop them. To best protect expensive electronic components and appliances against damaging voltage surges, you need a good grounding system combined with surge protection at both the main panel and at the appliance (for more information on the grounding system, see Chapter 2).

Through induction

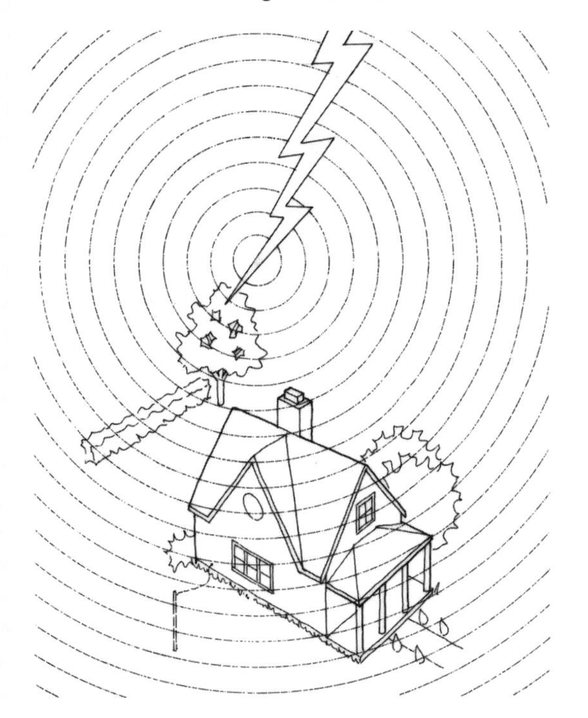

When lightning flashes across the sky, it creates magnetic lines of force. These forces surge into the house, inducing voltages into anything that can conduct electricity.

Through power lines

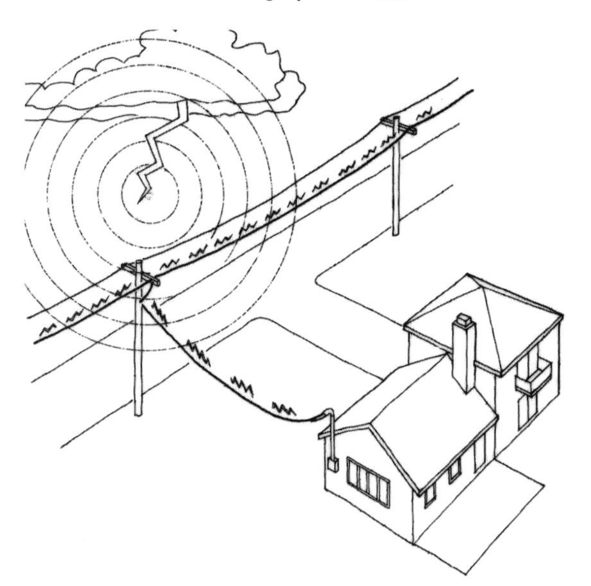

Lightning can induce a voltage surge into the high-voltage power lines. The surge travels along the lines, into the service drop, and into the house.

luckily, did not hit him. Most lightning surges are less dramatic, but that doesn't mean less damaging to appliances.

Smaller voltage surges are caused by appliances within the home, such as switches, motors, solenoids, and compressors turning on and off. These small surges occasionally damage other components or put annoying noises into them. For instance, it is very common for a light switch turning on and off to put a noise spike large enough to send a horizontal line across the TV screen and to put a loud pop into stereo gear (even a boom box).

To protect against unwanted voltages, it's a good idea to install surge arresters, which deflect excess voltages away from electronic equipment and into the grounding system. But before installing the surge arresters, make sure your house has a good grounding system (for more on the grounding system, see Chapter 2). If the grounding system isn't in good condition, the arresters won't work properly.

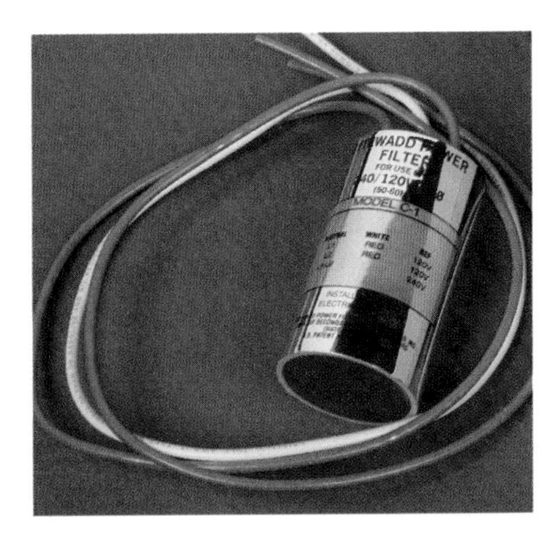

Hard-wired surge arresters, like the Tytewadd shown here, are installed at the main panel. They deflect voltage surges from the power lines harmlessly into the grounding system.

I always provide protection in two stages: at the main panel and then at the appliance (called point-of-use protection).

Main panel

A hard-wired surge arrester (see the photo below) installed at the main panel protects all the house wiring from major surges coming in on the power line. This type of arrester costs $150 and more at electrical supply stores.

A hard-wired surge arrester monitors the voltage coming into the house and deflects any voltages that exceed the standard house voltage to ground. Because this type of arrester is at the main panel, it should be installed by a qualified, licensed electrician.

Point of use

Point-of-use surge arresters deflect surge voltages back into the grounding system and are built with sensitive electronic components inside that eliminate noise from other appliances.

Point-of-use surge arresters are normally not sufficient on their own to protect against massive surges caused by lightning. Instead, they are a secondary defense, deflecting surges that get past the hard-wired surge arrester at the main panel.

There are two point-of-use designs: One is mounted directly onto a receptacle, and the other is a strip type, which most people are familiar with. Most computer stores, hardware stores, and even some discount stores sell point-of-use surge arresters. It doesn't pay to skimp here. The cheap units sold at most stores are light duty at best. It's better to spend more to get the maximum protection for your electronic components.

The unit I prefer is made by Isobar: the model Isotel Ultra 6 (see the photo at right). It comes with a $25,000 insurance policy if equipment plugged into this unit is damaged by a voltage surge.

There are three reasons why I like this particular model. First, it tells you if there is any problem with the house wiring, just like a plug-in receptacle tester would. Second, it is heavy duty, so it can deflect massive voltage surges to the grounding system. Third, it has three banks of two receptacles (six total), with each bank being electronically isolated from the others. That means any component plugged into a receptacle in one bank will not transmit noise into a component plugged into another bank. It also has an outlet for the telephone line to protect phones, faxes, modems, and answering systems.

An Isobar surge arrester will cost around $100. If you think the price is high, think about the cost of replacing a television or stereo that gets destroyed by a voltage surge.

This point-of-use surge arrester comes with a $25,000 insurance policy, has six receptacles that are electronically isolated from one another, and protects telephone lines. (Photo courtesy of Isobar.)

9

SPECIAL INSTALLATIONS

As our homes become more and more appliance oriented, more folks are attempting installations themselves. There are good and bad sides to this story. On the good side, by installing and wiring appliances, a homeowner can save significant money over hiring an electrician to do the job. On the bad side, many appliances draw a lot of voltage, so they have different installation requirements than standard 120-volt circuits.

It is also common for a modern home to require telephone jacks in almost every room, for hooking up to phones, fax machines, computer modems, and even satellite dishes (see Chapter 8). Because the homeowner is responsible for installing and maintaining the house telephone system, many are opting to run the cables themselves, instead of hiring a contractor.

Whether wiring an appliance or the phone system, it's important to do the job correctly the first time around to avoid problems and expensive service calls. In this chapter, I'll show you how to wire some of the most common household appliances, and I'll also illustrate how to install telephone wiring in the home.

WIRING APPLIANCES

An appliance is either hard wired or plugged into a receptacle. When a large appliance is going to be permanently installed at one location, it is normally hard wired. This means that the incoming cable that powers the appliance is brought into the unit and is spliced to its internal wiring. This is done in a splice box on the appliance supplied by the manufacturer. Large appliances that have a cord and plug include electric ovens and dryers.

Here I'll illustrate how to wire dishwashers, in-wall electric ovens and drop-in cooktops, garbage disposals, baseboard heaters, water heaters, electric ovens, and electric dryers. Most of the appliances covered in this chapter are required by the National Electric Code (NEC) or the manufacturer to be on a dedicated circuit (for more on dedicated circuits, see the sidebar on the facing page). And according to code, all appliances require a disconnect means in the form of a switch with a clearly defined "off" position that opens all hot conductors or an accessible plug and cord. In all cases, you should make sure power is off before wiring, and you must always follow the manufacturer's installation instructions.

Dedicated Circuits

A dedicated circuit is one that supplies power to one specific appliance or to a receptacle or receptacles in a specific room or area. Nothing else can be fed off this circuit. It is assumed that the load will need most, if not all, of the power provided by the circuit. If some of the power is drawn elsewhere, it could adversely affect the load on the circuit.

I made a service call in which the homeowner said the motor on the water pump was intermittently overheating and just didn't sound right. I found that the owner had tapped the pump circuit into the water-heater circuit. Whenever the water heater turned on, it drew power from the pump, causing the pump to work harder and eventually overheat.

The National Electric Code (NEC) and appliance manufacturers dictate when and where dedicated circuits are required. In reference to general-purpose receptacles, according to the NEC, no one cord-and-plug appliance (portable) can pull more than 50% of the branch-circuit rating on a circuit that has lights or other appliances. That means 7.5 amps is the maximum load for one such appliance on a 15-amp circuit, and 10 amps is the maximum load for a 20-amp circuit. If that load is exceeded, put the appliance on a dedicated circuit.

If an appliance on a dedicated circuit has a cord and plug, the receptacle must be located as close as possible to the appliance and no farther than 6 ft. You are not allowed to plug in the appliance to the receptacle using an extension cord.

Also, the NEC and appliance manufacturers require heavy-duty appliances, such as dryers and water heaters, to be on dedicated circuits. And if an appliance manufacturer requests that the appliance be on a dedicated circuit, the NEC requires you to follow the instructions (it will be indicated in the literature supplied with the appliance).

Dedicated heavy-duty loads for appliances that run continuously for three hours or more at a time, such as electric baseboard heaters, should never exceed 80% of the branch-circuit rating. That means 12 amps is the maximum a load or group of loads can pull on a 15-amp circuit, and 16 amps is the maximum for a 20-amp circuit.

A dedicated circuit can also be used to isolate one area electrically from another. For instance, in the kitchen, the light over the sink cannot be wired off the countertop receptacle circuit because the light could drain valuable power from the receptacles, which are dedicated to countertop appliances.

Another reason for wanting to isolate one area from another is to keep the noise of one electronic device from affecting another. For instance, a light switch can sometimes put out a noise spike large enough that it can be heard on the stereo and could interfere with the TV picture. I always put the living-room receptacles on a dedicated circuit to isolate them from noisy switch circuits.

WHERE TO INSTALL DEDICATED CIRCUITS

The following is a small list of specific areas and heavy-duty appliances that require or that should have dedicated circuits.

- Baseboard heaters
- Bathroom fans with built-in heaters
- Bathroom receptacles
- Dishwashers
- Dryers
- Electric ranges and ovens
- Garage receptacles
- Garbage disposals
- Kitchen/dining/pantry receptacles
- Large microwave ovens
- Large portable air conditioners
- Large stereo systems
- Laundry-room receptacles
- Shop receptacles
- Water heaters (including small under-sink models)
- Water pumps
- Welders

There are many designs of each appliance in this chapter, so I'm not going to discuss which models are best. I'm going to focus on the wiring only. The most sound buying advice I can offer is to do some research to find the best one for your situation. Shop around, ask a lot of questions, and check consumer magazines for comparisons and other information.

Dishwasher

The most common hard-wired appliance in the kitchen is the dishwasher. It is typically 120 volts and requires a dedicated 20-amp circuit using 12-gauge wire.

The first step is to turn off power to the circuit. Then bring the cable through the floor or wall anywhere behind the dishwasher (for more on running cable, see Chapter 3). Leave enough slack in the cable so that the dishwasher can be pulled out and serviced—usually 3 ft. to 4 ft. of slack will be sufficient.

The dishwasher will have a small metal splice box in front (behind the kick panel) to house the wire splices. Before bringing the cable into the box, remove the box's cover and install an NM connector in the knockout. The NM connector will prevent

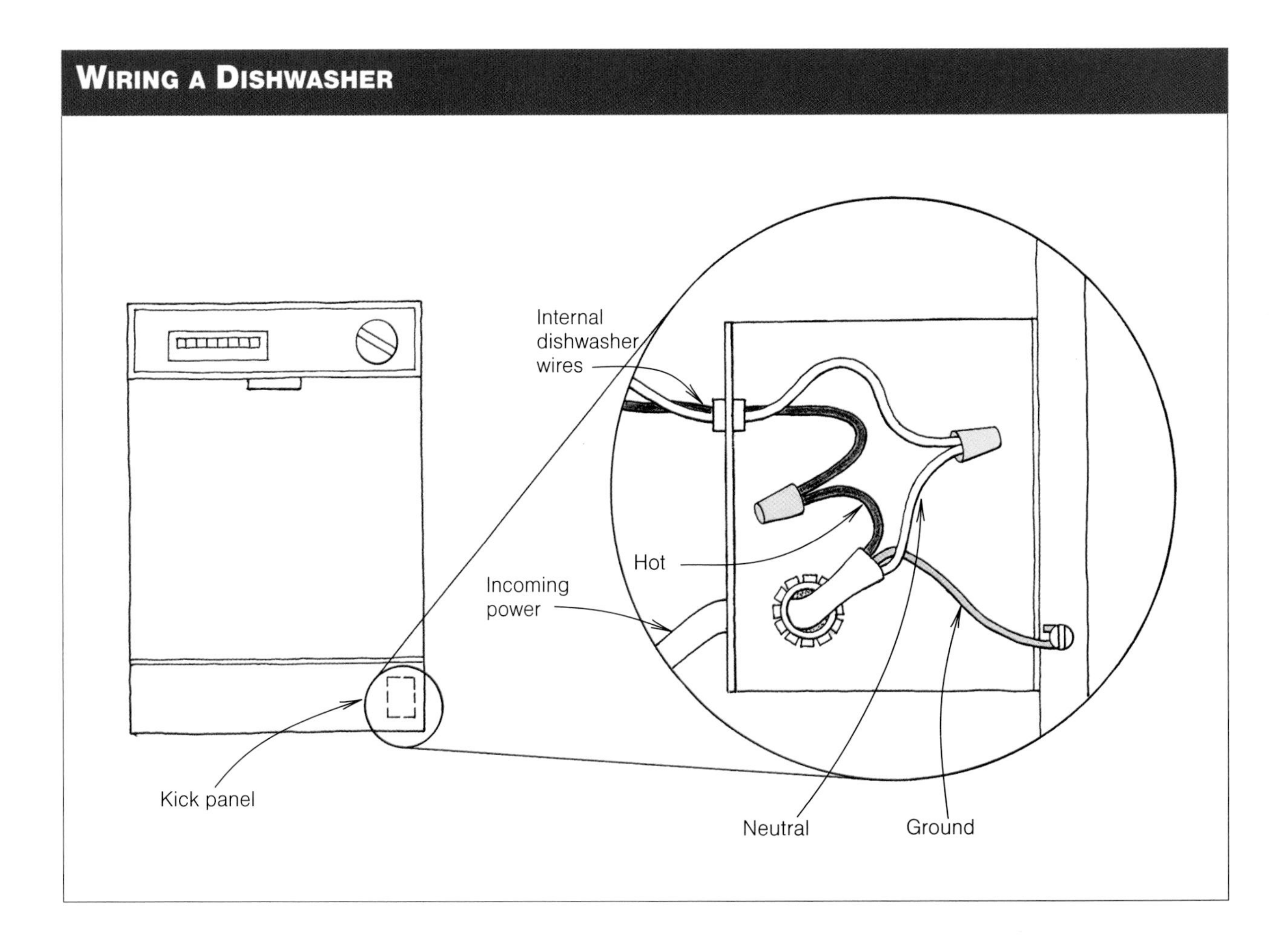

the cable from being cut by the sharp metal edge of the box, which can happen easily since all dishwashers vibrate.

Bring the cable into the splice box, strip it, and make the connections (for more on stripping cable, see p. 36). Connect the bare copper wire to the green grounding screw. Splice the incoming black wire to the internal black wire, and splice the white neutrals together (see the drawing on the facing page). Twist the wire splices together and cover them with wire nuts (for more on splicing, see p. 40). Be sure no bare wire is showing under any wire nut, then gently fold the wires into the splice box, replace the cover and kick panel, restore power, and the dishwasher is ready to run.

In-wall electric oven and drop-in cooktop

An in-wall electric oven and drop-in cooktop are normally put on a hard-wired dedicated circuit. Because they use both 120 and 240 volts (the bake unit and burners use 240, and the lights, buzzers, and timers use 120), the incoming cable has four conductors: one insulated black wire (hot), one insulated red wire (hot), one ground, and one insulated white wire (neutral).

Wiring an In-Wall Electric Oven and Drop-In Cooktop

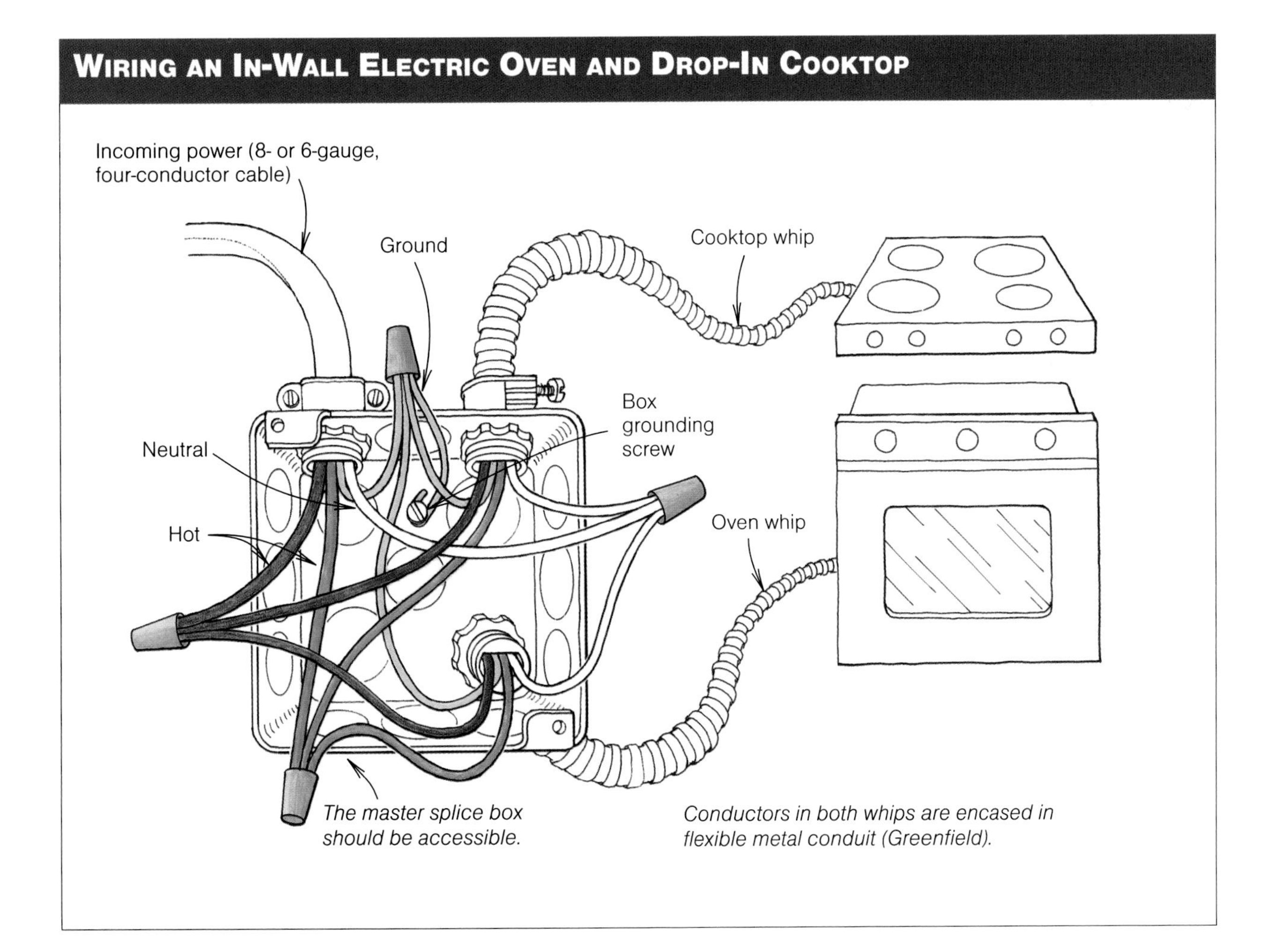

The incoming power cable is normally either 8-gauge or 6-gauge copper. With 8-gauge conductors, you'll need 40-amp overcurrent protection, and with 6-gauge conductors, you'll need 50-amp overcurrent protection. Both the oven and cooktop will come with a whip, which is a pigtail of wires encased in a flexible metal conduit (Greenfield). The wires within the whip are usually smaller gauge (10, 12, or 14 gauge) than the feeder cable (the one from the main panel).

To wire the oven and cooktop, first turn off the power. Then install a large metal splice box—42 cu. in. (4 11/16 in. square and 2 1/8 in. deep). Because the splice box is a maintainable item (according to the NEC), it must remain accessible, so don't hide it behind a wall. Instead, put it under a countertop so that it's out of view but accessible.

Bring the incoming power cable and both whips to the splice box (for more on running cables, see Chapter 3). Because the splice box is metal, you should run all the cables through NM connectors to prevent the sharp edges of the box from cutting through them.

Once you have both whips and the incoming cable in the box, make the connections (see the drawing on p. 129). Simply splice like wires—black to black, red to red, white to white, and ground to ground. You'll also need to ground the metal splice box by running a pigtail from the box to the ground splice. Then simply fold the wires back into the box, put the cover plate on, restore power, and you're ready to cook.

Garbage disposal

A garbage disposal can be plug and cord, but it is normally hard wired. It uses 120 volts, and often local codes require you to put the disposal on a dedicated circuit. This is a good idea because the disposal can generate a lot of noise in an electrical circuit.

Follow the manufacturer's instructions for installation of the unit. The installation procedures may vary, depending on the manufacturer, but the wiring is basically the same. First turn off the power. Once the disposal has been installed, remove the cover plate at the bottom of the unit to expose the internal wires.

Install and wire a single-pole switch in the wall immediately above the sink (for more on installing switches, see Chapter 3), then run a cable from the switch to the garbage disposal. Although there's no official requirement or stapling schedule, I hold the cable near the center of the stud with three staples to keep the cable far enough from the finished wall so that errant screws and nails won't damage it. Many local codes require the cable to be protected in Greenfield (flexible metal conduit) where it is exposed under the sink. You should check your local codes to see if this is a requirement in your town.

Once the switch has been hooked up, and the cable has been run to the disposal, simply connect like wires under the cover plate of the disposal (see the drawing on the facing page): black to black and white to white (the ground connects to a grounding screw on the unit). Restore power, and the unit is ready to run.

Wiring a Garbage Disposal

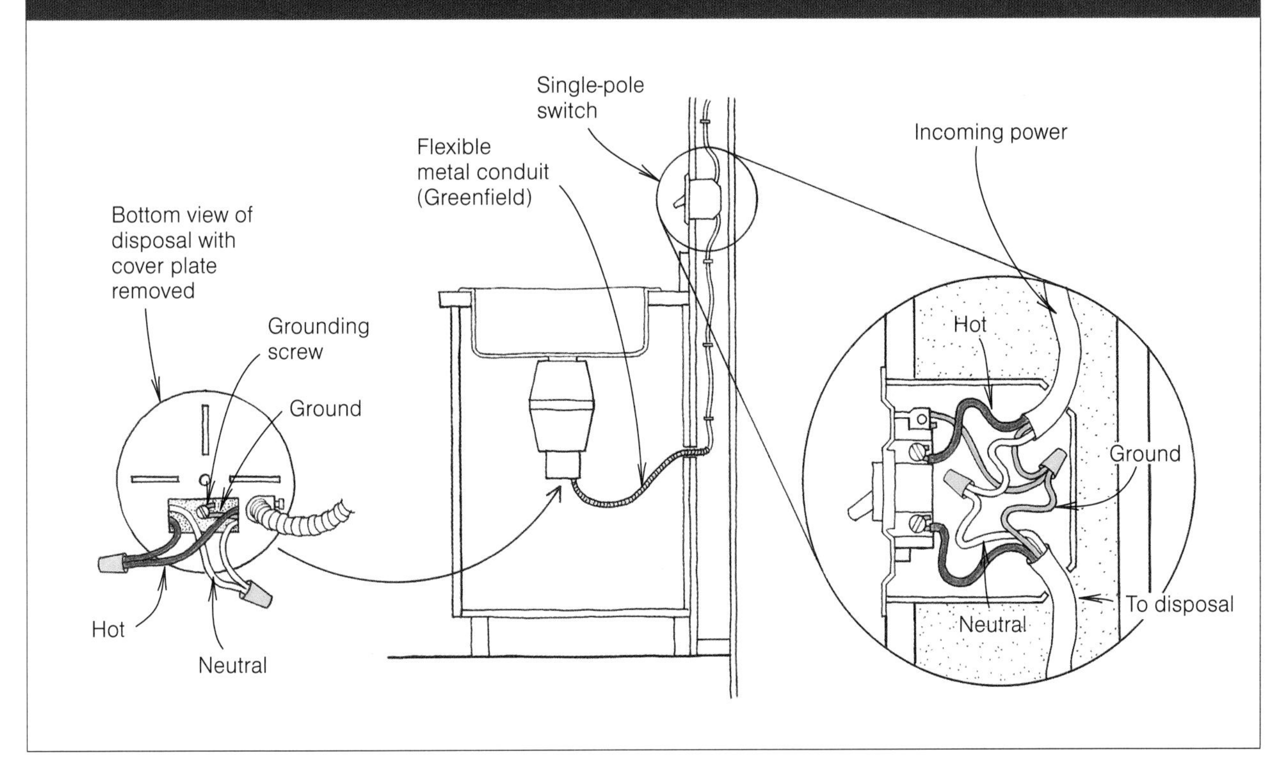

Electric baseboard heater

A 240-volt electric baseboard heater provides fast heat and is installed easily. Because a baseboard heater pulls a lot of current, the manufacturer typically requires that it be put on a dedicated circuit, and the NEC requires you to follow the manufacturer's instructions. Although there are 120-volt baseboard heaters available, I don't recommend them because they pull twice as much current as the 240-volt units, so they are less efficient. Also, they could be accidentally hooked up to a 240-volt circuit, which will burn out the heating elements inside and could eventually start a fire. This discussion applies to 240-volt installations only.

What size cable you use to feed the heater depends on the total amperage of the unit. A baseboard heater pulls about 1 amp per ft., so to figure out the total amperage of the unit, simply add up the total length of the heater. For instance, if you want to install a 10-ft. unit, it will pull about 10 amps. Use 12-gauge cable if the heater will pull less than 16 amps and 10-gauge cable if the heater will pull above 16 amps but less than 24 amps.

Installation of a baseboard heater is pretty simple, although procedures may vary, depending on the manufacturer, so be sure to follow the manufacturer's guidelines. The biggest question is whether you want to mount the thermostat in the wall or in the heater. Which method you choose will affect where you bring the incoming power cable to.

Wiring a wall-mount thermostat

Although you are allowed to replace an existing single-pole thermostat with one of the same, in a new installation, a wall-mount thermostat for a baseboard heater is required by the NEC to be double pole because it is safer.

You see, a single-pole thermostat cannot remove all power to the heater. It only disconnects one of the two 240-volt legs of the circuit (a 240-volt-only appliance requires two hot wires, called legs, and no neutral). A double-pole thermostat, on the other hand, has an off position with two disconnects that allow you to remove power from both 240-volt legs of the circuit. Because many people will have an existing single-pole thermostat for an electric baseboard heater, I have shown the wiring for it in the top drawing on the facing page to make replacing one easier. But because of the code regulation, the discussion that follows is for a double-pole thermostat.

To install a double-pole thermostat, first be sure power to the circuit is turned off. Then run the cable from the main panel to where the thermostat will be located. Run another cable from the thermostat to the heater. The cable must be brought to one end of the heater, where the connections will be made.

The instructions will tell you the mounting height of the thermostat. It can be installed in a standard receptacle or switch box, but be sure the box has at least 23 cu. in. of volume (3½ in. deep) because a double-pole thermostat needs lots of wiring space.

A double-pole thermostat has a line side and a load side. Connect the incoming power to the line side and the outgoing power to the load side (see the bottom drawing on the facing page). Connect the ground wire to the grounding terminal. Because there will be two hot legs on the circuit, there will be no neutral, and the white wire will be hot. The NEC requires this wire to be taped black to indicate it's hot.

Once the connections have been made at the thermostat, you can make the connections at the heater. A standard baseboard heater can be connected to the thermostat through either end. Simply remove the cover to access the wiring. Inside you will see two wires spliced with a wire nut (see the drawing on p. 134). Remove the wire nut and pull the wires apart (leave the wires on the other end of the heater alone).

Now pull the cable into the unit (install an NM connector in the knockout) and connect the incoming white hot wire (tape it black) to one of the two wires under the wire nut and the incoming black hot wire to the other wire (polarity doesn't matter here, so you don't have to worry about which wires the incoming wires are connected to). Then connect the ground wire to the grounding terminal. With that done, replace the cover.

Before you power up the heater, however, double-check the wiring to the thermostat. A very common mistake is to wire the incoming hot wires immediately across the thermostat switch (one to the line side and the other to the load side). When the thermostat is turned on, a

Wiring Wall-Mount Thermostats for Baseboard Heaters

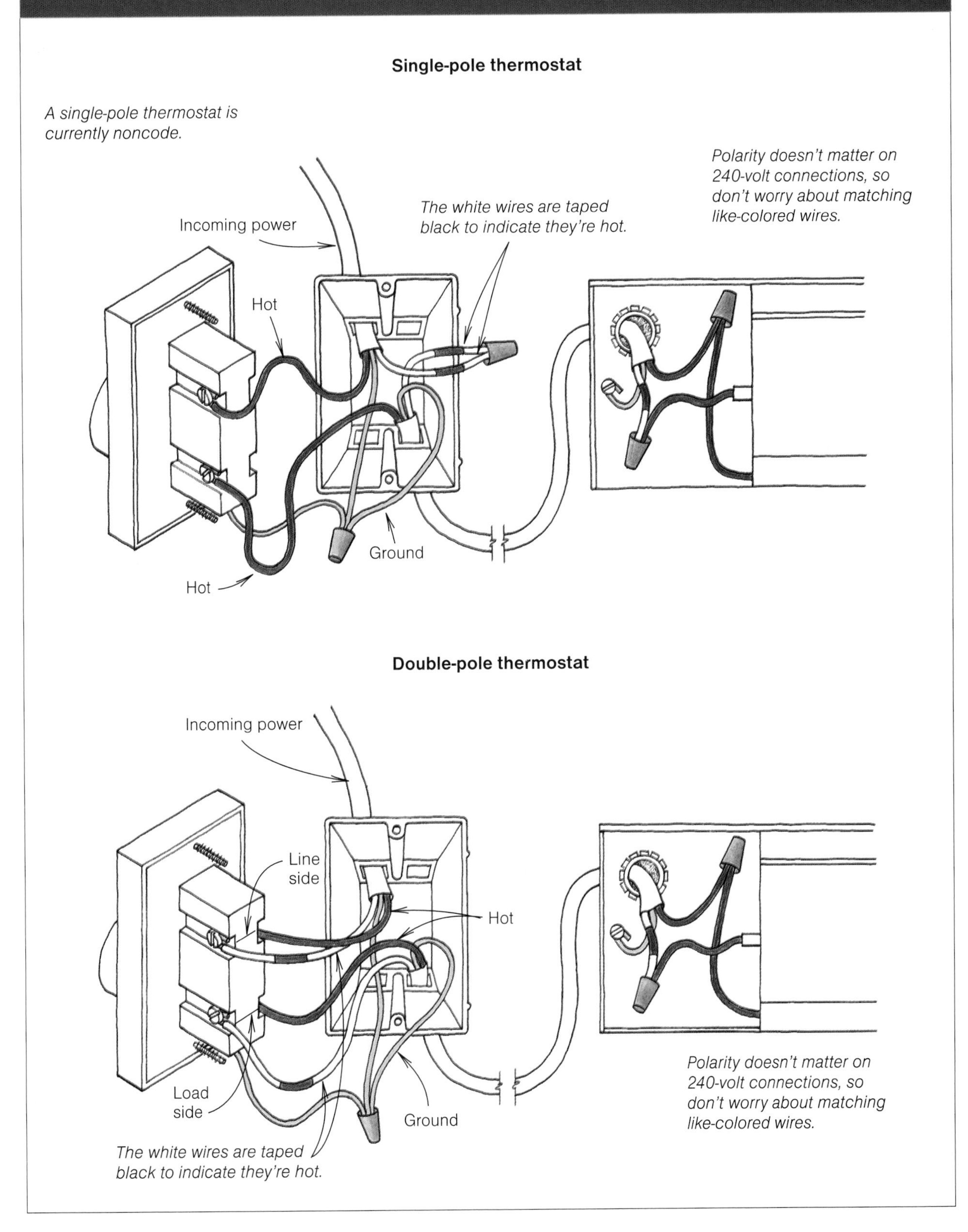

Preparing a Baseboard Heater for Wiring

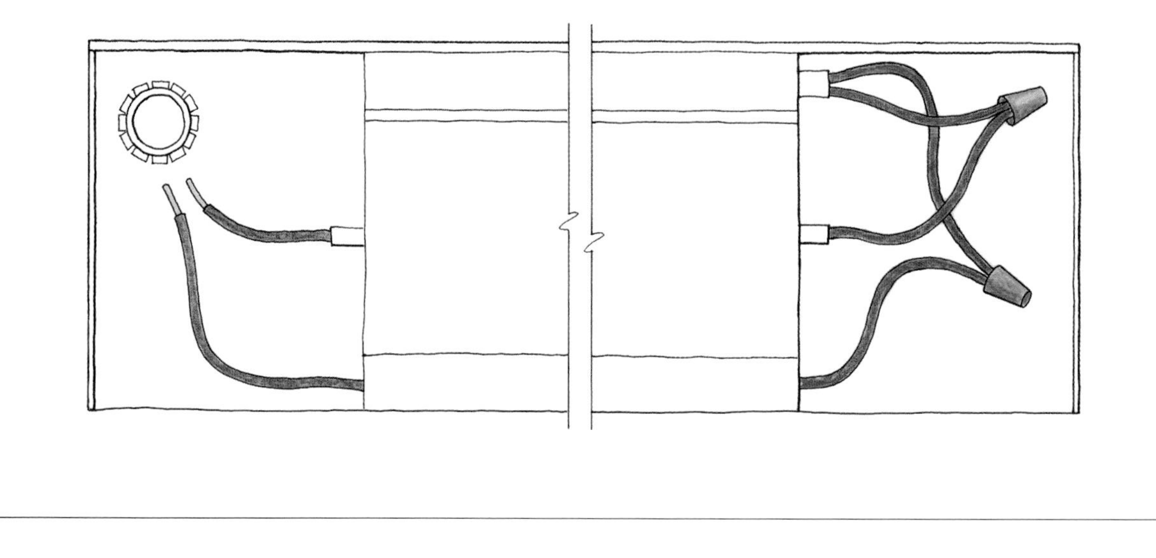

direct short will occur across these wires. Make sure both hot feeder wires are attached to the line side. If you've reversed these wires and turn on the heater, you'll hear a loud pop from the thermostat, and the breaker will trip off. After double-checking the wiring, restore power and test the unit.

When you turn on an electric baseboard heater for the first time, it is quite normal for the unit to smoke a little and to smell. But don't worry. It's just the element being broken in. You may want to open the doors and windows while you break in the unit. A heater may also smoke and smell after sitting idle for a long period of time.

Wiring an in-heater thermostat

Putting the thermostat in the end of the baseboard heater is easier than installing a wall-mounted thermostat because you simply run one cable to the heater's location. An in-heater thermostat can be either single or double pole (most are single pole).

The first step is to pull the cable from the main panel to one end of the heater. Once you have the cable to the correct location, install the heater following the manufacturer's instructions.

Before wiring, make sure power is turned off. Then remove the cover from one end of the unit and untwist the wires, leaving the other end alone (see the drawing above). Install an NM connector in the unit's knockout to protect the cable from the sharp edges. Pull the cable through and make the connections. Again, this is a 240-volt circuit, so both the black and white wires will be hot (tape the white wire black to indicate it's hot).

Wiring In-Heater Thermostats

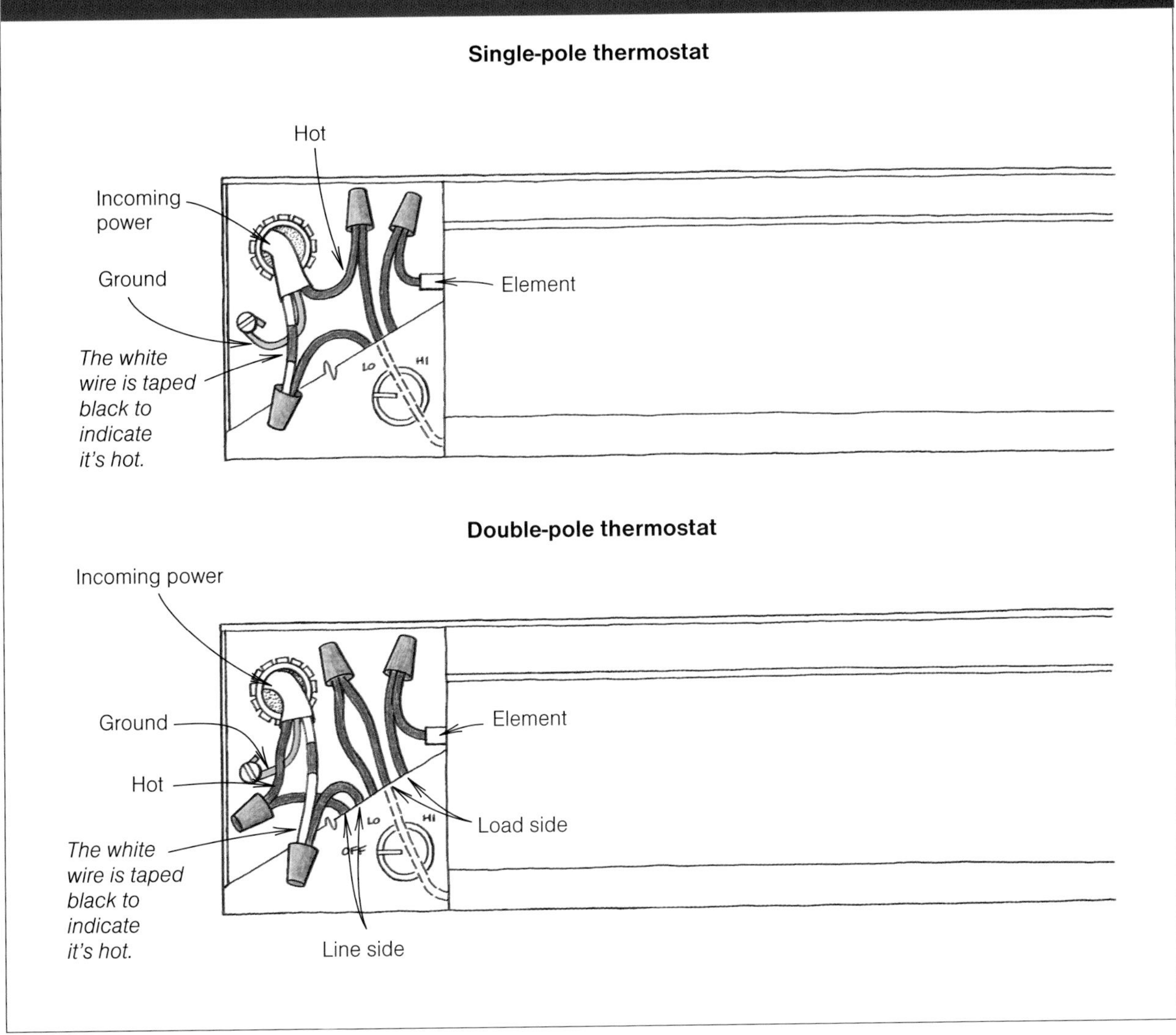

For a single-pole thermostat, connect one incoming hot wire to the thermostat and the other incoming hot wire to the element (there's no polarity involved because this is a 240-volt circuit). Then connect the ground wire to the grounding terminal (see the top drawing above).

For a double-pole thermostat, connect the incoming hot wires to the line side of the thermostat (see the bottom drawing above). Then connect the load side of the thermostat to the heater. Again, polarity doesn't matter here. Once the wiring has been done, restore power and test the heater.

Water heater

Wiring an electric water heater can be done by most anyone, but it's important that you follow all the manufacturer's installation instructions for safety and for optimum performance.

Wiring a Water Heater

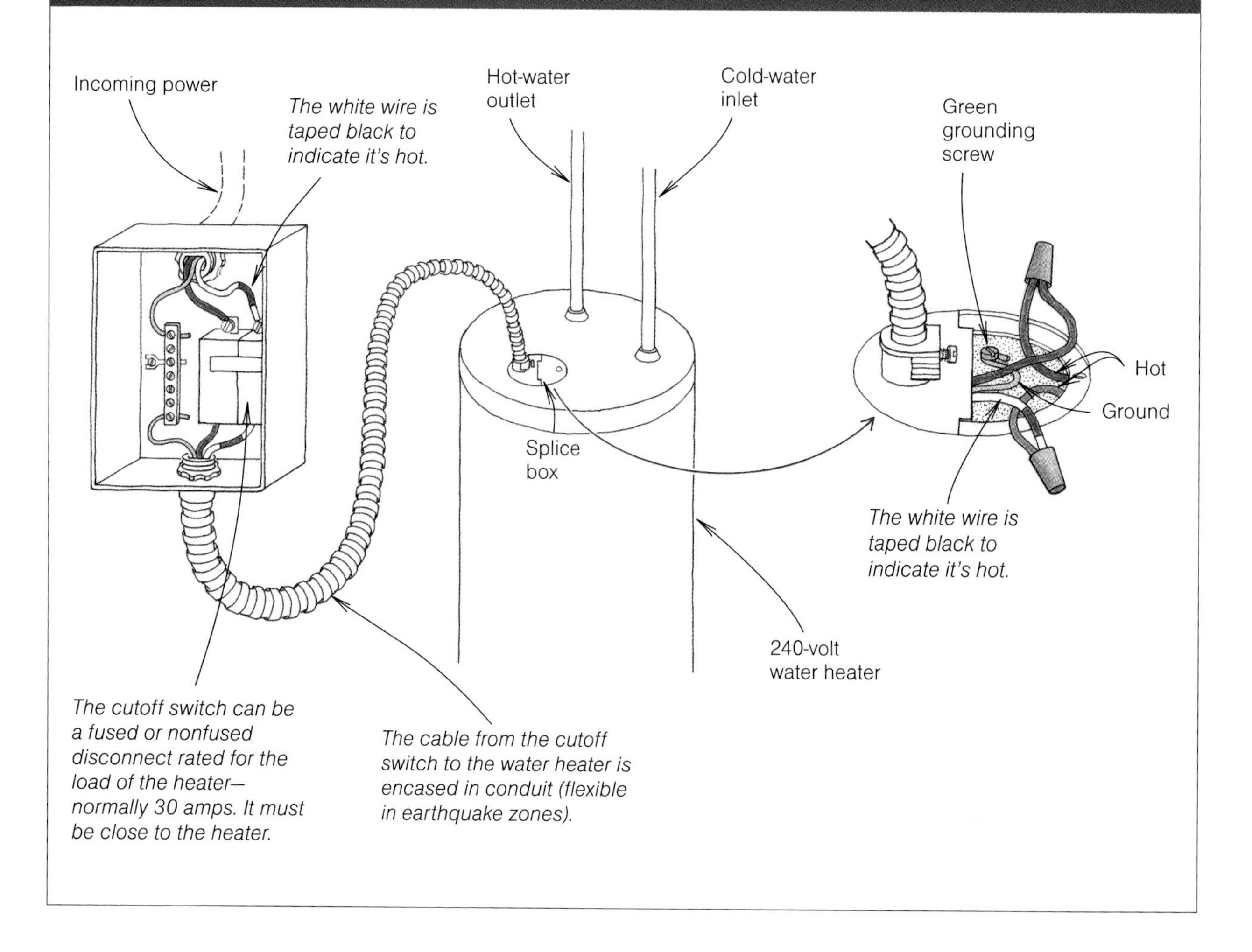

A standard water heater is comprised of an outer jacket, a layer of insulation, and a water tank. Depending on the maker of the unit, the heater will have either one or two heating elements that heat the water. The elements are controlled by a thermostat (one for each) and an internal overload switch, which cuts off power if the unit overheats.

An electric water heater is required by the manufacturer to be on a hard-wired dedicated circuit, and the NEC says you must follow the manufacturer's guidelines. Although some smaller water heaters (30 gal.) use 120 volts, the standard 50-gal. electric water heater installed today is 240 volts with a 4,500-watt heating element (some models have two elements). Because the element pulls a lot of amps, it should be wired with 10-gauge cable and should be protected by a 30-amp fuse or breaker.

If you are replacing a water heater, and the existing wiring is 12 gauge, you can still install the standard 4,500-watt heater. However, you'll have to use a smaller element, such as 3,500 watts (elements are available in wattages from 1,500 to 4,500) so that you do not overheat the wires. There is nothing wrong with using a smaller-wattage element in a larger heater: It just takes longer to heat the water.

But because the water heater has a label on its jacket that lists the specs of the unit (including the element wattage), there is a possible safety risk involved with installing a smaller-wattage element. If the element blows at any point, a plumber, electrician, or do-it-yourselfer will look at that label to find the size of the replacement element. If he installs the 4,500-watt element noted on the label, and the heater has 12-gauge wires, the element could overheat the wires. To avoid any confusion, if you replace the standard element with a lower-wattage one, indicate that you have done so on the water heater's label—and use large letters.

A cutoff switch (also called a disconnect switch or lockout) is required if the water heater is out of sight of the main panel. Its purpose is to provide a measure of safety while you work on the heater: No one can mistakenly turn on the power without you knowing about it. The cutoff switch can be a fused or nonfused disconnect (either a breaker or double-pole switch) rated for 240 volts.

Bring the cable from the main panel to the cutoff-switch box, which should be located as close as possible to the heater. Install NM connectors in the knockouts to protect the cable. Then pull the cable into the box and connect the incoming wires to the incoming terminals on the switch.

Run another cable from the cutoff switch to the water heater's splice box (typically located on top of the unit). If the heater is in a habitable area, such as a basement, the cable from the cutoff switch to the heater must be enclosed in conduit (flexible in earthquake zones).

Connect the cable to the outgoing terminals on the cutoff switch. At the heater's splice box, connect the two incoming hot wires to the two hot wires of the heater—no polarity is required, so don't worry about matching the colors of the wires. Assuming the heater pulls 240 volts, the white wire will be hot, so cover it with black tape to indicate it's hot. Do so at the heater, at the cutoff switch, and at the main panel.

Be sure to ground the water heater via the grounding terminal in the splice box. An ungrounded water heater is a safety hazard. I had one service call in which the homeowner complained of getting shocked off the copper plumbing lines as he was trying to repair a split where the metal pipes had frozen. Sometimes he would receive a shock, and other times he wouldn't. I traced the fault current back to the water heater, and I noticed that the water heater was ungrounded. I pulled off the cover over the bottom element and saw that the element had corroded. Its wire (with screw attached) had pulled away and was touching the metal case of the heater. Every time the upper thermostat applied power to the bottom element, the current would short over to the metal jacket and into the metal plumbing pipes. A very dangerous situation indeed.

If you have metal water pipes, it's also important to run a bonding jumper around the water heater, as shown in the drawing on p. 21. The bonding jumper will allow the ground connection to remain intact even if the water heater is nonmetallic or if it is removed for maintenance or is replaced.

Testing the water heater Once all the wiring has been completed, make the plumbing connections. There will be a cold-water inlet and a hot-water outlet at the top of the heater.

After the plumbing connections have been made, test the water heater. But it's important to fill the heater completely with water before restoring power to the circuit. If there is an air pocket inside the heater when the unit is turned on, an element could blow.

To fill the tank, turn on the cold-water valve on the cold-water inlet, then turn on a hot-water faucet in the house. When water flows continuously out of the hot-water faucet, the tank is full. Once the tank is full, restore power, turn on the water heater, and you're ready for a hot shower in around an hour.

Electric oven

A standard slide-in electric oven uses 120 volts and 240 volts (120 for the timer and 240 for the burners and bake unit) and plugs into a 50-amp receptacle wired with 6-gauge copper, four-conductor cable (two hot wires, one neutral, and one ground). It must be on a dedicated circuit and requires 40- or 50-amp overcurrent protection.

The receptacle must be mounted low so that the bottom panel or drawer of the oven can be removed (in theory) to disconnect the unit. If you don't have access to the receptacle, the NEC requires you to wire it through a cutoff switch installed in an accessible location.

The receptacle can be surface mounted or flush mounted. I find that the surface mount is easier to install because it comes with its own housing, or box. First, turn off the power. Then bring the cable to the location, if it's not already there, and screw the receptacle to the base of the wall (to a stud or to the floor). Then terminate the wires under the appropriate set screws on the receptacles (see the top drawing on the facing page), which will be marked for proper wiring. After wiring, attach the cover, restore power, plug in the oven, and turn it on.

Installation of a flush-mount receptacle is more difficult, so it lends itself to new construction or installations in which the wall is already opened. A flush-mount receptacle requires a large-volume box (30 cu. in.) that is sold separately, as well as a special cover plate to which the receptacle is mounted. First, make sure power is off, then securely attach the box to a stud and bring the cable into it. Put the cover plate on the box and make the connections to the receptacle (see the bottom drawing on the facing page). Again, the receptacle terminals will be labeled, so you'll know which wires go where. Once wired, attach the receptacle to the cover plate, restore power, and plug in the oven.

Electric dryer

A standard electric dryer is a cord-and-plug appliance that uses both 120 and 240 volts (120 for the timer and 240 for the heating element).

The drawings on p. 140 show three methods of wiring a dryer receptacle. The two on the left are in violation of the NEC because the ground becomes a current-carrying conductor—it is attached to the neutral terminal of the receptacle.

Wiring Oven Receptacles

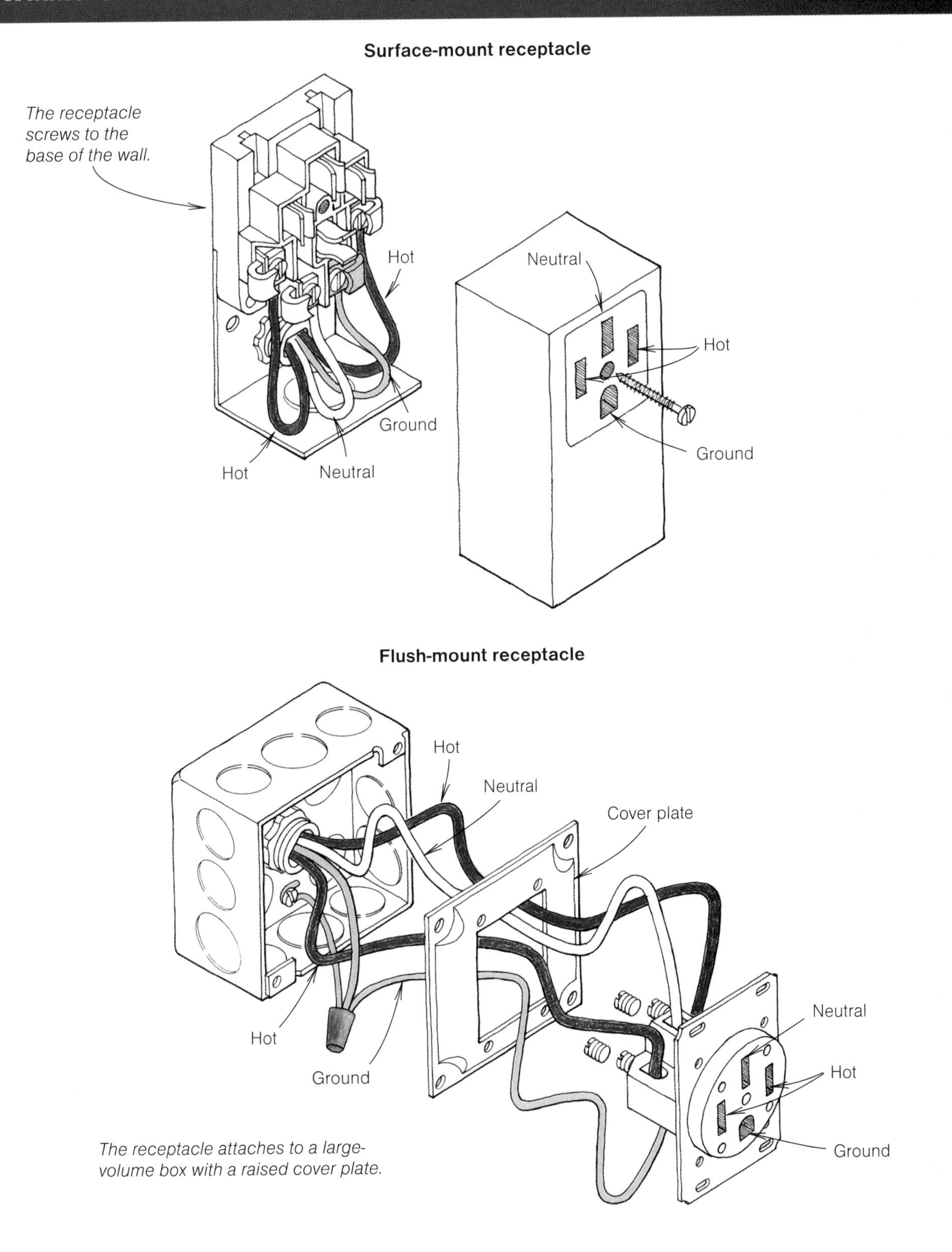

Wiring a Dryer Receptacle

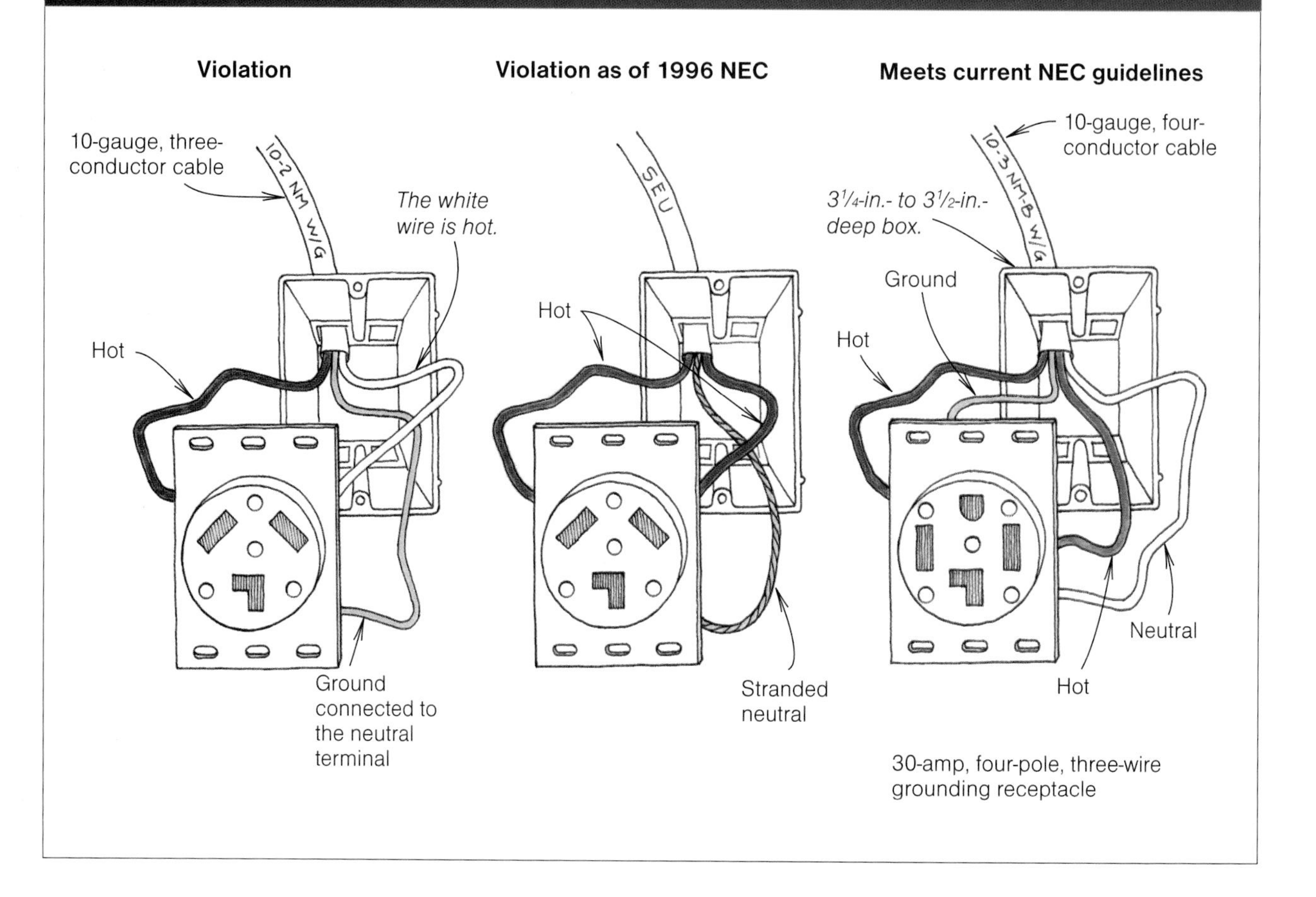

Anyone touching the ground downstream could be electrocuted. If you have either of these, I recommend replacing them with a code-approved receptacle and cable (see the right drawing above).

The NEC now requires the dryer to have a four-blade plug and the receptacle to be wired with 10-gauge, four-conductor cable—two hot wires, one neutral, and one ground—with 30-amp overcurrent protection. A dryer receptacle is 30 amps and can be surface mounted or flush mounted, and the installation procedures are pretty much the same as those for an electric-oven receptacle.

A surface-mount receptacle comes with its own housing, or box. To install it, cut off power, bring the cable to the location, and attach the receptacle to the wall. Then terminate the wires under the appropriate set screws on the receptacle, which will be marked for proper wiring. After wiring, attach the cover, restore power, plug in the dryer, and turn it on.

A flush-mount dryer receptacle requires a large-volume box (30 cu. in.) that will be sold separately, as well as a special cover plate to which the receptacle is mounted. Simply follow the same steps you did for a flush-mount oven receptacle. Turn off power, then securely attach the box to a stud within the wall and bring the cable into it. Put the cover plate on the box and make the connections to the receptacle. As with the surface-mount receptacle, the terminals will be labeled. Once wired, attach the receptacle to the cover plate, restore power, and plug in the dryer.

Troubleshooting

All appliances come with an instruction manual complete with troubleshooting steps that will help you locate and solve sources of difficulties with the appliance, so there's no need to spend a lot of time on this process here. But there are a few general wiring mistakes to look for—before calling the maintenance man—that will make an appliance malfunction or not work at all.

If the appliance is not working at all, check the splices and the connections (turn off power for this) to be sure they are still tight. If the appliance has a cord and plug, make sure the cord has not worked loose. If the connections seem okay, make sure the breaker or fuse has not tripped off.

If the appliance is working, but the breaker or fuse continues to trip off when the appliance is turned on, a likely cause of the problem is that the wrong-size wire gauge was used in the installation. Check the manufacturer's instructions to see that the proper-gauge wire and overprotection (fuse and breaker size) have been used. For instance, if the manufacturer recommends 12-gauge wire and 20-amp protection, be sure the installer hasn't used 14-gauge wire and 15-amp protection (for more on wire gauge and overcurrent protection, see Chapter 2). If you have moved into a new home, it's a good idea to check the wire size to be sure it's the correct gauge for the appliance. Also see if the appliance is supposed to be on a dedicated circuit. Other loads or appliances placed on the circuit could be causing an overload, which will make the overcurrent-protection device cut off power to the circuit.

If all these things turn out to be okay, the wires could be damaged or too old and worn, causing a short somewhere along the run, which will trip the breaker or blow the fuse. To check the wiring, remove power from the circuit and disconnect the appliance, capping the hot wire with a wire nut so that it can't short out against anything metal (you can simply unplug a cord-and-plug appliance). Then restore power to the circuit. If the breaker trips off or the fuse blows, the wires are bad and should be replaced. If nothing happens, the problem lies with the appliance and you should call a maintenance person. You can do this test for any appliance.

TELEPHONE WIRING

The homeowner is now responsible for the in-house telephone system, which means that either you do the wiring and maintain it or you hire a contractor. Because phone wiring is low voltage and the wires are small, which makes them easy to pull through walls, there's no reason why anyone can't handle this job.

In this section I will tell you how to install telephone wiring properly so that it will be easy to maintain and will provide many years of good service. There are two steps to the installation process: connecting to the protector box where the telephone-company cable enters the house, and wiring and installing the wall-mounted phone jacks.

Connecting to the protector box

The protector box, officially called a network interface device (NID), is located on the outside of the house usually near the electrical meter base. It provides the interface between the telephone-company cable and the in-house telephone wiring. The telephone-company cable stops here and the in-house wiring begins.

Although different protector boxes may be wired differently, the incoming cable from the phone company usually enters the bottom left of the box and feeds two modular telephone jacks in top. The jacks allow for the wiring of two phone lines—two different numbers—in the house. The jacks at the top of the box are wired into four small screw terminals on the right-hand side of the protector box. The top two terminals are for one line (one phone number), and the bottom two are for a separate phone line. Any number of house telephone wires can connect to these little terminals. (Unlike electrical installations, it's okay to put a number of wires under each terminal.)

Typical home telephone cable is composed of four wires called "two pair." One pair of wires is red and green, and the other is black and yellow. The red and green pair is the primary pair; the black and yellow pair is the backup. The backup pair is needed in case the first pair gets damaged, and it is also used when you want to add a second phone to that line. Multipair cable (which has 4 to 12 pairs of wires) is available for feeding multiple phone lines, but these are not typical installations. Phone cable can be obtained from your local telephone company, hardware store, or electrical supply store.

To connect the in-house phone cable to the protector box, bring the cable in through the grommet on the bottom right of the box. Once inside, run the cable to the terminals through the gutter in the right side of the box. Remove about ½ in. of insulation from each wire end and connect the red wire to one terminal and the green to the other terminal. Each terminal will have a red or green wire attached to it already, so you just have to match like-colored wires. Then fold the backup pair away from the terminals for possible future use.

If you want to feed multiple phone jacks from these terminals, and you can't fit all the wires under the terminals, take a multipair cable from the protector box to a master splice box inside the house. (The master splice box is nothing more than a standard double-gang nonmetallic electrical box.) From the master splice box, you can wire as many jacks as you need to and splice the cables to the multipair cable. You can splice telephone wires by twisting them together, by twisting them together and then soldering them, by tightening them under screw terminals, or by using a mechanical splice device called a Scotchlock (made by 3M and available at your local phone store or electrical supply store).

Installing phone jacks

Telephone cable is run through the house just as electrical cable is run (see Chapter 3). But because it's thinner and more flexible, pulling it through the walls is easier. But be careful. Because the cable is thinner, it's also easier to break. You can also run the cable outside the walls along the baseboard trim. This type of run is perfect for add-ons.

Once you get the cable to the proper location, install the wall jack. You have two options. If the wires are inside the wall, you can install a standard single-gang electrical box in the wall and attach the jack to that. If the wires have been run outside the walls, you can buy a special jack that attaches to baseboard trim. The special jack can be installed without cutting a large hole in the wall for the box. Phone jacks can be purchased just about anywhere, including your local telephone store and hardware store.

Wiring Telephone Jacks

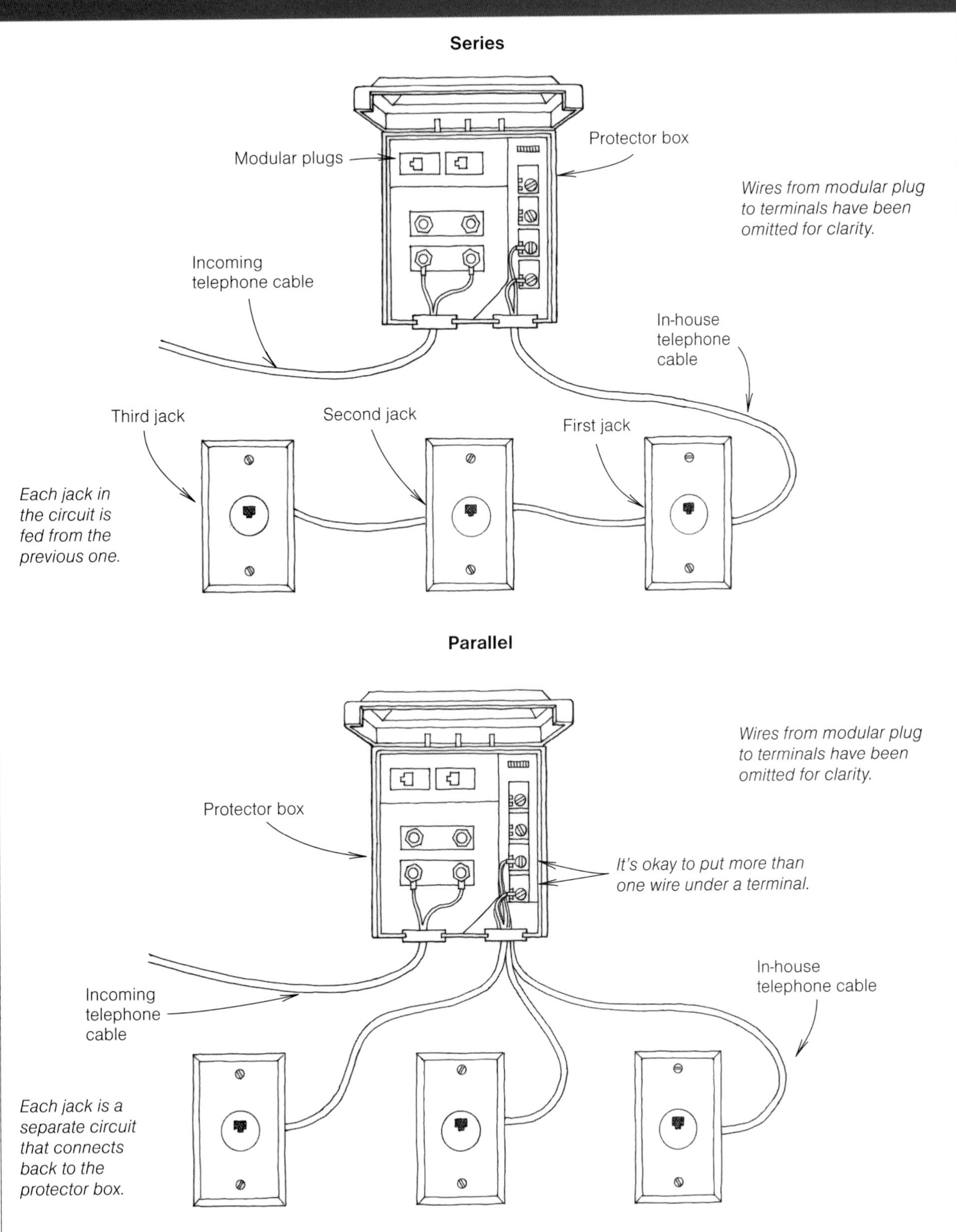

Wiring phone jacks

There are two basic systems of wiring wall-mounted phone jacks in a house: series or parallel (see the drawings on p. 143). The theory behind each system is identical to receptacles wired in series or parallel (see Chapter 3).

Wiring in series is not preferred here because if one break occurs along the line, everything after that break is disconnected. It also takes longer to troubleshoot a faulty jack because the problem could be with any of the jacks or wires in the circuit. Unfortunately, unless you ask otherwise, series probably is what you'll get even if you hire a professional to run the cables and install the wall jacks. To wire in series, connect the cable to the terminals in the protector box, and feed one wall jack. The first jack will feed the second, the second will feed the third, and so on.

Wiring in parallel is the only way I go because troubleshooting is easier. If one jack is not working, I know the problem lies in that one circuit: I don't have to check every one to find the trouble. Wiring in parallel takes a little longer and is slightly more expensive, but it's worth it. To wire in parallel, run a separate cable for each jack you are installing. If you can't fit all the wires under the terminals, run a multipair cable from the protector box to a master splice box and begin the runs from there.

Regardless of whether you wire in series or parallel, the connections at the phone jack are identical. Strip about 2 in. of sheathing from the cable end and then strip about ½ in. from the ends of the red and green wires. Wind the red wire around the terminal that already has a red wire around it, and do the same with the green wire. Fold the yellow and black wires back for possible use later.

Once the attachments have been made, put the wall jack in place. If you have an outlet box with a standard wall jack, attach the jack to the box with the screws provided. If you have used a jack that mounts outside the wall, screw it to the baseboard trim or to the wall.

Adding a jack

If you want to add a jack to the phone system, you can simply run a new cable back to the protector box or to the master splice box from the new jack (wiring in parallel), or you can feed it off an existing jack (wiring in series). The latter is not preferred, but it will be easier because you won't have to cut through the walls. Instead, you can simply run the cable along the baseboard trim and connect to a special surface-mount phone jack.

Other options for adding phones are to replace a single wall jack with a double jack or to add a modular Y adapter to the single jack—both will provide two connections at one location. Then you can simply buy two phone cables with plugs on either end and run them along the base of the wall and to both phones.

Troubleshooting

There are two problems that are typical for phone lines: no dial tone and noise on the line. If you are not getting a dial tone from your phone lines, you must first find out if the trouble lies outside or within the house. As I said before, the phone company is responsible only for the wiring to the protector box from the outside.

Bring a phone outside to the protector box and pull the incoming telephone cable out of the box's jack (top left of the box). It pushes in and pulls out like a standard modular cord. Plug your phone into the jack and listen for a dial tone (wait two minutes for the telephone equipment to reset). If you get one, the problem is in your house. If not, the problem is with the wiring to the house, so call the phone company.

If the problem is with the in-house wiring, how you proceed will depend on whether the system was wired in series or in parallel. In a series system, if one jack loses the dial tone, all the others after it will lose the signal. The trick is finding which jack has the problem. If all the jacks are not working, the trouble lies in the connection from the protector box to the first one. Unfortunately, you have to figure out which jack is the first by tracing the cable from the protector box (which may or may not be possible to do).

If the system is wired in parallel, only one jack will not be working—the others will still work. First remove the wall jack that's not working. Check for burned or corroded terminals on the jack. If you see any, replace the jack. If the terminals are in good shape, check the condition of the wires and make sure the connections are tight. If all seems okay at the jack, follow the same steps at the protector box outside the house.

If both locations look to be in good shape, remove the red and green wires at the jack and protector box and use the backup pair (the black and yellow wires). If you still don't get a dial tone, the cable is probably broken along its run or the jack is bad. Try replacing the jack first. If that doesn't work, run a new cable.

If you have a lot of noise on the line, the source could be one of the wires shorting out on a piece of metal (typical if you used a metal outlet box for the jack) or just a loose, corroded, or burned connection at the jack. Remove the jack from the wall and inspect the wires. If you see any bare spots, tape them up. If the jack has loose or corroded terminals, replace it. Also make sure that all connections are tight. If none of these solves the problem, try switching the cable pair and using the backup pair.

Another noise problem common on phone lines is a phenomenon called cross talk, which comes from the telephone company. Somewhere in the main telephone cables, a signal on a cable pair adjacent to your phone cable has a signal that is too loud and is feeding into your cable pair, allowing you to hear talking on another line.

Cross talk normally happens when radio stations broadcast their transmissions over local telephone lines to a remote transmitter. They jack the signal up higher than they're supposed to so they can overcome signal loss as it travels to the remote transmitter. The telephone company may say the problem doesn't exist. If you don't get satisfactory answers from the telephone company about cross talk, file a complaint with the State Corporation Commission. The State Corporation Commission is an agency responsible for enforcement of the duties of all public service companies, including phone and utility companies. You'll find the number in the government pages of your phone book.

GLOSSARY

Alternating current (AC) Current that first flows in one direction and then returns. It creates a sine wave that rises above and below a 0-voltage reference point.

Ampacity The current-carrying capacity of a wire. For example, the ampacity of a 12-gauge wire is 20 amps.

Ampere (amp) The unit of measure for current flow. The amperage of a circuit can be measured with a multimeter. See also Current.

Arcing A luminous discharge of electrical current between two points.

Armored cable (BX) A flexible, metal-clad cable with the wires factory-installed.

Bonding Connecting two conductors mechanically to assure electrical continuity.

Branch circuit A wire or conductor that starts at the main panel and feeds one or more loads.

Cable Two or more conductors inside a sheath.

Capacitance The ability of an object to retain electrical energy.

Circuit A complete path of current between the power source and the load.

Circuit breaker A mechanical device that will open an electrical circuit when a predetermined amount of current is exceeded.

Conduit A plastic or metal pipe used to protect wires and cable. See also Armored cable and Greenfield.

Conductor Any material that can conduct electricity. Normally a copper or aluminum wire.

Continuity An uninterrupted electrical path between two points.

Current The organized flow of electrons from one point to another on a circuit. It is measured in amperes (amps). Current can be increased by raising the voltage or by lowering resistance. It can be decreased by lowering the voltage or by increasing resistance.

Cutoff switch A switch that can disconnect the electricity to an appliance during an emergency or for maintenance. It may or may not be fused. Also called a disconnect.

Dedicated circuit A circuit dedicated to one specific appliance(s) or to a receptacle or receptacles in one specific area.

Derate To lower the amount of current a circuit is allowed to have.

Direct current (DC) Current that flows in one direction only. Normally created from a chemical reaction within a battery or within a power supply of electronic equipment.

Disconnect See Cutoff switch.

Electrical bus See Grounding bus, Hot bus, and Neutral bus.

Equipment grounding conductor The conductor that connects the frame of an appliance into the grounding system. It is normally the bare wire in a cable.

Fuse A device that opens a circuit when a predetermined amount of electrical current is exceeded.

Gauge The diameter of a wire referenced to the American Wire Gauge Standard of measurement. The larger the diameter, the less resistance a conductor will have to current flow. In general, the larger the gauge number, the smaller the wire diameter. For example, 12-gauge wire is larger in diameter than 18-gauge wire.

Greenfield A flexible-metal wire casing installed in the field to protect exposed wires or cable.

Ground fault A short circuit that occurs when a hot wire touches a ground or a noncurrent-carrying conductor.

Ground-fault circuit interrupter (GFCI) A safety device that compares the amount of current going to a load to the amount coming back. It will open the circuit in a flash if the return current is lower because it means the current is following a path through something else (possibly a person).

Ground rod An 8-ft. to 10-ft. rod driven into the earth near the main panel to bleed stray grounding currents and surges away from the house and into the earth. The rod is normally ⅝-in.-dia. copper-clad or galvanized metal.

Ground wire The wire inside a cable (normally the bare wire) that connects to the receptacle or switch grounding screw. See also Equipment grounding conductor.

Grounded Connected directly or indirectly to earth. A piece of equipment is grounded if a grounding wire (equipment grounding conductor) connects it back to the neutral/grounding bus at the main panel.

Grounding bus See Neutral/grounding bus.

Grounding electrode The ground rod and anything else that the grounding electrode conductor connects into the earth.

Grounding electrode conductor A bare copper wire, usually 6 or 4 gauge, that connects the house electrical system to the ground rods.

Grounding system—Any part of the electrical system that directly or indirectly connects to earth ground.

Hot bus The flat, insulated metal bar that runs down the center of the main panel and connects to the incoming hot power cable.

Hot wire A wire that carries current to a load. It is normally a red or black wire.

Impedance The total sum of the opposition to current flow.

Induced voltage Voltage that is created when the lines of an electromagnetic field cut into a conductor.

Knockouts Removable sections of both metal and nonmetallic boxes (also on panel boxes) through which incoming and outgoing cables are passed.

Load The user of supplied electricity, such as a toaster, air conditioner, or hard-wired appliance.

Main breaker The large circuit breaker in the top of the main panel that all of the house current flows through (also called the main). It will disconnect all the power to the house if a predetermined amount of current is exceeded.

Main panel The main fuse box or circuit-breaker box where the utility wires from the meter base come to and where all the branch-circuit wires originate. Also called the panel board, service panel, or just panel.

Multimeter A piece of test equipment used to measure voltage, current, and resistance. (Some will also perform a continuity test.)

National Electric Code (NEC) The organization that sets and publishes voluntary nationwide wiring codes, which are adopted and enforced by official agencies to keep electrical wiring and installation of that wiring standardized and safe. The code is revised every three years.

Neutral The return path of AC current to its source. See also Neutral current.

Neutral/grounding bus The metal bus within the main panel that the neutrals (white wires) of the branch-circuit cables connect to. In a main panel the neutral bus is more correctly called a neutral grounding bus because it has all the bare grounding wires connected to it as well and is also connected to earth ground via the ground rods.

Neutral current The return current from a load to the main panel and ultimately to the utility transformer.

Neutral wire The white insulated wire in a circuit between a load and the main panel providing a path for return, or neutral, current.

Noncurrent-carrying conductor A material that can conduct electricity but normally does not, such as the metal frame of an appliance.

Ohm Unit of measure for resistance.

Overload Occurs when a load is placed on a circuit that exceeds the rating of the wiring and the circuit breaker or fuse.

Panel bond In reference to the main panel, the attachment of the panel's neutral/grounding bus to the metal framework of the panel via a screw or wire. If a hot wire hits a bonded panel, the breaker belonging to the hot wire will trip. If the panel is not bonded, the panel frame would remain at the same potential as the hot wire hitting it and would shock anyone who touched it. The same logic would apply to the frame of any metal-panel housing such as the cut-off panel for a water heater.

Panel box The metal box that holds the circuit breakers or fuses.

Pigtail A small wire approximately 6 in. long.

Polarity A reference to polarization, meaning a specific wire must connect to a specific terminal.

Power Normally measured in watts, it is the amount of electricity you are actually using or referencing too.

Primary voltage The voltage on the utility side of the transformer. Normally 7,200 volts or 19,900 volts.

Protector box A panel installed outside the house that acts as the interface between the incoming telephone cable and the house telephone lines.

Resistance The opposition to current flow. The unit of measure for resistance is ohms and is designated by the word or the Greek omega sign (Ω). Resistance can be measured with a multimeter.

Secondary voltage The utility voltage on the house side of the utility transformer, normally 120/240 volts.

Service entrance The path the service-entrance cable takes between the utility transformer and the main panel.

Service-entrance cable The large-diameter cable that brings the secondary utility voltage to the main panel.

Short circuit Occurs when hot and neutral wires touch, making the neutral wire current carrying.

Signal gain The signal-gathering ability of an antenna.

Sone A subjective unit of loudness equal to the intensity of a 1,000-hertz tone 40 decibels above the listener's own threshold of hearing.

Transformer A device used to transform electrical energy from one circuit to another. It can lower the energy or increase it. A transformer at the utility pole lowers the high voltage on the power lines to a voltage that can be used in the house. Transformers at electrical substations increase the voltage so that it doesn't dissipate over long distances.

UL approved A product that is approved by United Laboratories for its intended use.

UL listed A product that is on a list for approval by United Laboratories.

United Laboratories (UL) An independent testing laboratory that sets standards for products and certifies that those products comply with those standards. In simple terms, UL verifies that a manufacturer's product will do what the manufacturer says it will do. A lot of electrical equipment requires a label indicating that UL has tested the equipment for its intended use, or it may not pass inspection.

Watt Unit of energy that indicates how much power an appliance will use. See Power.

Volt Unit of measure of electrical pressure (voltage).

Voltage Electrical pressure created by a chemical reaction, as in a car battery, or mechanically created by a generator. It is measured in volts, and when a quantity of voltage is written, a capital V or the word volts normally follows the numerical quantity. Voltage can be measured with a multimeter.

Voltage drop The amount of voltage lost along a conductor from the power source to the load. In general, the longer the distance between power source and load, the greater the voltage drop.

Voltage surge A voltage spike that exceeds the 120-volt line voltage. Typically cause by lightning strikes or electrical malfunctions.

Volt-ohm meter (VOM) The predecessor to the multimeter.

Wire nut Insulated mechanical wire connector originated by the Ideal company.

Yoke The metal support around a switch or receptacle that holds it together and fastens it to the outlet box.

INDEX

PUBLISHER: **Jon Miller**

ACQUISITIONS EDITOR: **Julie Trelstad**

ASSISTANT EDITOR: **Karen Liljedahl**

EDITOR: **Thomas McKenna**

LAYOUT ARTIST: **Amy Bernard**

PHOTOGRAPHER, EXCEPT WHERE NOTED: **Susan Kahn**

ILLUSTRATOR: **Vincent Babak**

TYPEFACE: **Frutiger**

PAPER: **70-lb. Mead Moistrite Matte**

PRINTER: **Quebecor Printing/Hawkins, New Canton, Tennessee**